MATLAB®

An Introduction with Applications

Second Edition

Amos Gilat

Department of Mechanical Engineering
The Ohio State University

JOHN WILEY & SONS, INC.

ACQUISITIONS EDITOR	Joseph Hayton
SENIOR MARKETING MANAGER	Jennifer Powers
SENIOR PRODUCTION EDITOR	Ken Santor
COVER DESIGNER	Kevin Murphy

This book was set in Times New Roman by the author and printed and bound by Malloy, Inc. The cover was printed by Lehigh Press.

This book is printed on acid free paper. ∞

ISBN 0-471-69420-7

Printed in the United States of America

10 9 8 7 6 5

Preface

MATLAB® is a very popular language for technical computing used by students, engineers, and scientists in universities, research institutes, and industries all over the world. The software is popular because it is powerful and easy to use. For university freshmen in it can be thought of as the next tool to use after the graphic calculator in high school.

This book was written following several years of teaching the software to freshmen in an introductory engineering course. The objective was to write a book that teaches the software in a friendly, non-intimidating fashion. Therefore, the book is written in simple and direct language. In many places bullets, rather than lengthy text, are being used to list facts and details that are related to a specific topic. The book includes numerous sample problems in mathematics, science, and engineering that are similar to problems encountered by new users of MATLAB.

This second edition of the book is updated for MATLAB 7. It includes also a new chapter (11) about symbolic math operations with MATLAB, and a new section in Chapter 4 that shows how to import and export data.

I would like to thank several of my colleagues at The Ohio State University. Professors Richard Freuler, Mark Walter, Brian Harper, and Walter Lampert, and Dr. Mike Parke for reading sections of the book and suggesting modifications. I also appreciate the involvement and support of Professors Robert Gustafson and John Demel and Dr. John Merrill from the First-Year Engineering Program at The Ohio State University. Special thanks to Professor Mike Lichtensteiger (OSU), and my daughter Tal Gilat (Stanford University), who carefully reviewed the entire book and provided valuable comments and criticisms.

I would like to express my appreciation to all those who have reviewed this text at its various stages of development, including Betty Barr, University of Houston; Andrei G. Chakhovskoi, University of California, Davis; Roger King, University of Toledo; Richard Kwor, University of Colorado at Colorado Springs; Larry Lagerstrom, University of California, Davis; Yueh-Jaw Lin, University of Akron; H. David Sheets, Canisius College; Geb Thomas, University of Iowa; Brian Vick, Virginia Polytechnic Institute and State University; Jay Weitzen, University of Massachusetts, Lowell; and Jane Patterson Fife, The Ohio State University. In addition, I would like to acknowledge the support of Joe Hayton, Ken Santor, Caroline Sieg, Katherine Hepburn, Simon Durkin, John Stout, and Jay Beck, all from John Wiley & Sons.

I hope that the book will be useful and will help the users of MATLAB to enjoy the software.

Amos Gilat
Columbus, Ohio
May, 2004

To my parents Schoschana and Haim Gelbwacks

Contents

Chapter 9   Three-Dimensional Plots   239

Chapter 10   Applications in Numerical Analysis   261

Chapter 11   Symbolic Math   283

Introduction

MATLAB is a powerful language for technical computing. The name MATLAB stands for MATrix LABoratory, because its basic data element is a matrix (array). MATLAB can be used for math computations, modeling and simulations, data analysis and processing, visualization and graphics, and algorithm development.

MATLAB is widely used in universities and colleges in introductory and advanced courses in mathematics, science, and especially in engineering. In industry the software is used in research, development and design. The standard MATLAB program has tools (functions) that can be used to solve common problems. In addition, MATLAB has optional toolboxes that are a collection of specialized programs designed to solve specific types of problems. Examples include toolboxes for signal processing, symbolic calculations, and control systems.

Until recently, most of the users of MATLAB have been people who had previous knowledge of programming languages such as FORTRAN or C, and switched to MATLAB as the software became popular. Consequently, the majority of the literature that has been written about MATLAB assumes that the reader has knowledge of computer programming. Books about MATLAB often address advanced topics, or applications, that are specialized to a particular field. In the last few years, however, MATLAB is being introduced to college students as the first (and sometimes the only) computer program they learn. For these students there is a need for a book that teaches MATLAB assuming no prior experience in computer programming.

The Purpose of this Book

MATLAB: An Introduction with Applications is intended for students who are using MATLAB for the first time and have little or no experience in computer programming. It can be used as a textbook in freshmen engineering courses, or workshops where MATLAB is being taught. The book can also serve as a reference in more advanced science and engineering courses when MATLAB is used as a tool for solving problems. It also can be used for self study of MATLAB by students and practicing engineers. In addition, the book can be a supplement or a secondary book in courses where MATLAB is used, but the instructor does not have the time to cover it extensively.

Topics Covered

MATLAB is a huge program, therefore it is impossible to cover all of it in one book. This book focuses primarily on the foundations of MATLAB. It is believed that once these foundations are well understood, the student will be able to learn advanced topics easily by using the information in the Help menu.

The order in which the topics are presented in this book was chosen carefully, based on several years of experience in teaching MATLAB in an introductory engineering course. The topics are presented in an order that allows the student to follow the book chapter after chapter. Every topic is presented completely in one place and then is used in the following chapters.

The first chapter describes the basic structure and features of MATLAB and how to use the program for simple arithmetic operations with scalars as with a calculator. The next two chapters are devoted to the topic of arrays. MATLAB's basic data element is an array that does not require dimensioning. This concept, which makes MATLAB a very powerful program, can be a little difficult to grasp for students that have only limited knowledge and experience with linear algebra and vector analysis. The book is written such that the concept of arrays is introduced gradually and then explained in extensive detail. Chapter 2 describes how to create arrays, and Chapter 3 covers mathematical operations with arrays.

Following the basics, script files are presented next in Chapter 4, followed by two-dimensional plotting in Chapter 5. The next topic covered, in Chapter 6, is function files, which is intentionally separated from the subject of script files. This has been proven to be easier to understand by students who are not familiar with similar concepts from other computer programs. Programming with MATLAB is covered in Chapter 7, which includes flow control with conditional statements and loops.

The next three chapters cover more advanced topics. Chapter 8 describes how MATLAB can be used for carrying out calculations with polynomials, and how to use MATLAB for curve fitting and interpolation. Plotting three-dimensional plots, which is an extension of the chapter on two-dimensional plots, is covered in Chapter 9. Chapter 10 covers applications of MATLAB for numerical analysis. It includes solving nonlinear equations, finding a minimum or a maximum of a function, numerical integration, and solution of first order ordinary differential equations. Chapter 11 is a new chapter that was added to the second edition of the book. It covers in great detail how to use MATLAB in symbolic operations.

The Framework of a Typical Chapter

In every chapter the topics are introduced gradually in an order that makes the concepts easy to understand. The use of MATLAB is demonstrated extensively within the text and by examples. Some of the longer examples in Chapters 1–3 are titled as tutorials. Every use of MATLAB is printed in the book with gray background. Additional explanations appear in boxed text with white background. The idea is that the reader will execute these demonstrations and tutorials in order to gain experience in using MATLAB. In addition, every chapter includes formal sample problems which are examples of applications of MATLAB for solving problems in math, science, and engineering. Each example includes a problem statement and a detailed solution. Some sample problems are presented in the middle of the chapter. All of the chapters (except Chapter 2) have a section at the end with several sample problems of applications. It should be pointed out that problems with MATLAB can be solved in many different ways. The solutions of the sample problems are written such that they are easy to follow. This means that in many cases the problem can be solved by writing a shorter, or sometimes "trickier," program. The students are encouraged to try to write their own solutions and compare the end results. At the end of each chapter there is a set of homework problems. They include general problems from math and science and problems from different disciplines of engineering.

Symbolic Calculations

MATLAB is essentially a software for numerical calculations. Symbolic math operations, however, can be executed if the Symbolic Math toolbox is installed. The Symbolic Math toolbox is included in the student version of the software and can be added to the standard program.

Software and Hardware

The MATLAB program, like most other software, is continually being developed and new versions are released frequently. This book covers MATLAB, Version 7, Release 14. It should be emphasized, however, that this book covers the basics of MATLAB which do not change that much from version to version. The book covers the use of MATLAB on computers that use the Windows operating system and almost everything is the same when MATLAB is used on other machines. The user is referred to the documentation of MATLAB for details on using MATLAB on other operating systems. It is assumed that the software is installed on the computer, and the user has basic knowledge of operating the computer.

The Order of Topics in the Book

It is probably impossible to write a textbook where all the subjects are presented in an order that is suitable for everyone. The order of topics in this book is such that the fundamentals of MATLAB are covered first (arrays and array operations), and, as mentioned before, every topic is covered completely in one location which makes the book easy to use as a reference. Some people, however, might want to follow a slightly different order, especially when the book is used as a text in a MATLAB class. For example, some teachers may cover the basics of script files (the first four sections of Chapter 4), before teaching Chapter 2 (creating arrays) and Chapter 3 (arrays operations). This will allow students to use script files instead of the Command Window when studying Chapters 2 and 3. Also, relational and logical operations (Section 7.1) and polynomials (Section 8.1) can be considered mathematical operations and presented together with the rest of the operations in Chapter 3.

Chapter 1

Starting with MATLAB

This chapter begins by describing the characteristics and purposes of the different windows in MATLAB. Next, the Command Window is introduced in detail and is the only one that is used in the rest of the chapter. Chapter 1 shows how to use MATLAB for arithmetic operations with scalars, similar to the way that a calculator is used. This includes the use of elementary math functions with scalars. The chapter then shows how to define scalar variables (the assignment operator) and how to use these variables in arithmetic calculations.

1.1 STARTING MATLAB, MATLAB WINDOWS

It is assumed that the software is installed on the computer, and that the user can start the program. Once the program starts, the window that opens, shown in Figure 1-1, contains three smaller windows which are the Command Window, the Current Directory Window, and the Command History Window. This is the default view of MATLAB. These windows are three of eight different windows in MATLAB. A list of the various windows and their purpose is given in Table 1-1. The **Start** button on the lower left side can be used to access MATLAB tools and features.

Four of the windows, the Command Window, the Figure Window, the Editor Window, and the Help Window, are used extensively throughout the book and are briefly described on the following page. More detailed descriptions are included in the chapters where they are used.

Command Window: The Command Window is MATLAB's main window, and opens when MATLAB is started. It is convenient to have the Command Window as the only visible window, and this can be done by either closing all the other windows (click on the **x** at the top right-hand side of the window you want to close), or by first selecting on the **Desktop Layout** in the **Desktop** manu, and then **Command Window Only** from the submenu that opens. How to work in the Command Window is described in detail in Section 1.2.

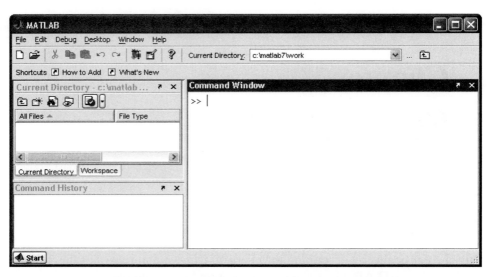

Figure 1-1: The default view of MATLAB desktop.

Table 1-1: MATLAB Windows

Window	Purpose
Command Window	Main window, enters variables, runs programs.
Figure Window	Contains output from graphic commands.
Editor Window	Creates and debugs script and function files.
Help Window	Provides help information.
Launch Pad Window	Provides access to tools, demos, and documentation.
Command History Window	Logs commands entered in the Command Window.
Workspace Window	Provides information about the variables that are used.
Current Directory Window	Shows the files in the current directory.

Figure Window: The Figure Window opens automatically when graphics commands are executed, and contains graphs created by these commands. An example of a Figure Window is shown in Figure 1-2. A more detailed description of this window is given in Chapter 5.

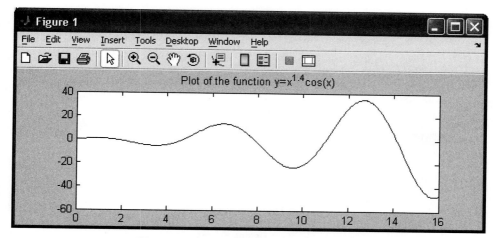

Figure 1-2: Example of a Figure Window.

Editor Window: The Editor Window is used for writing and editing programs. This window is opened from the **File** menu in the Command Window. An example of an Editor Window is shown in Figure 1-3. More details on the Editor Window are given in Chapter 4 where it is used for creating script files, and in Chapter 6 where it is used to create function files.

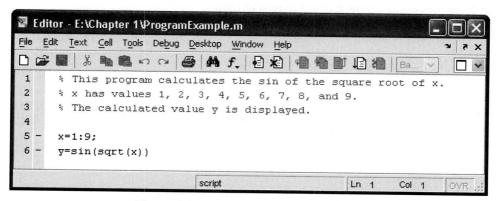

Figure 1-3: Example of an Editor Window.

Help Window: The Help Window contains help information. This window can be opened from the **Help** menu in the toolbar of any MATLAB window. The Help Window is interactive and can be used to obtain information on any feature of MATLAB. Figure 1-4 shows an open Help Window.

When MATLAB is started for the first time the screen looks like that shown in Figure 1-1 on page 6. For most beginners it is probably convenient to close all the windows except the Command Window. The closed windows can be reopened by selecting them from the **Desktop** menu. The windows shown in Figure 1-1 can be displayed by first selecting **Desktop Layout** in the **Desktop** menu and then **Default** from the submenu.

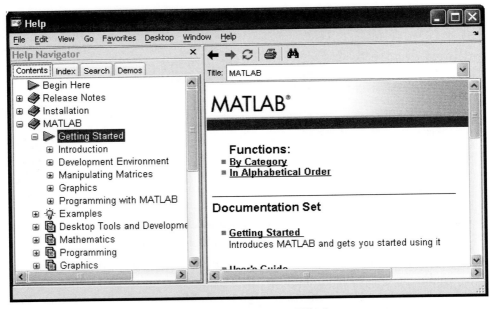

Figure 1-4: The Help Window.

1.2 WORKING IN THE COMMAND WINDOW

The Command Window is MATLAB's main window, and can be used for executing commands, opening other windows, running programs written by the user, and managing the software. An example of the Command Window, with several simple commands that will be explained later in this chapter, is shown in Figure 1-5.

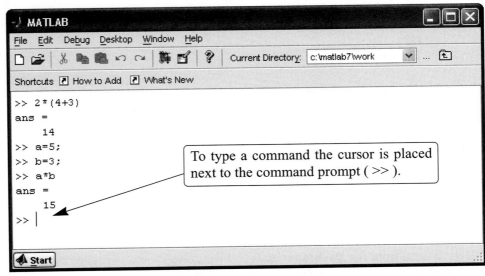

Figure 1-5: The Command Window.

Notes for working in the Command Window:

- To type a command the cursor must be placed next to the command prompt (>>).

- Once a command is typed and the **Enter** key is pressed, the command is executed. However, only the last command is executed. Everything executed previously is unchanged.

- Several commands can be typed in the same line. This is done by typing a comma between the commands. When the **Enter** key is pressed the commands are executed in order from left to right.

- It is not possible to go back to a previous line in the Command Window, make a correction, and then re-execute the command.

- A previously typed command can be recalled to the command prompt with the up-arrow key (↑). When the command is displayed at the command prompt, it can be modified if needed and executed. The down-arrow key (↓) can be used to move down the previously typed commands.

- If a command is too long to fit in one line, it can be continued to the next line by typing three periods … (called an ellipsis) and pressing the **Enter** key. The continuation of the command is then typed in the new line. The command can continue line after line up to a total of 4096 characters.

The semicolon (;):

When a command is typed in the Command Window and the **Enter** key is pressed, the command is executed. Any output that the command generates is displayed in the Command Window. If a semicolon (;) is typed at the end of a command the output of the command is not displayed. Typing a semicolon is useful when the result is obvious or known, or when the output is very large.

If several commands are typed in the same line, the output from any of the commands will not be displayed if a semicolon is typed between the commands instead of a comma.

Typing %:

When the symbol % (percent symbol) is typed in the beginning of a line, the line is designated as a comment. This means that when the **Enter** key is pressed the line is not executed. The % character followed by text (comment) can also be typed after a command (in the same line). This has no effect on the execution of the command.

Usually there is no need for comments in the Command Window. Comments, however, are frequently used in programs to add descriptions, or to explain the program (see Chapters 4 and 6).

The `clc` **command:**

The `clc` command (type `clc` and press **Enter**) clears the Command Window. After working in the Command Window for a while, the display may be very long. Once the `clc` command is executed a clear window is displayed. The command does not change anything that was done before. For example, if some variables were defined previously (see Section 1.6), they still exist and can be used. The up-arrow key can also be used to recall commands that were typed before.

1.3 ARITHMETIC OPERATIONS WITH SCALARS

In this chapter we discuss only arithmetic operations with scalars, which are numbers. As will be explained later in the chapter, numbers can be used in arithmetic calculations directly (as with a calculator), or they can be assigned to variables, which can subsequently be used in calculations. The symbols of arithmetic operations are:

Operation	Symbol	Example
Addition	+	$5 + 3$
Subtraction	–	$5 - 3$
Multiplication	*	$5 * 3$
Right division	/	$5 / 3$
Left division	\	$5 \backslash 3 = 3 / 5$
Exponentiation	^	$5 \wedge 3$ (means $5^3 = 125$)

It should be pointed out here that all the symbols except the left division are the same as in most calculators. For scalars, the left division is the inverse of the right division. The left division, however, is mostly used for operations with arrays, which are discussed in Chapter 3.

1.3.1 Order of Precedence

MATLAB executes the calculations according to the order of precedence displayed below. This order is the same as used in most calculators.

Precedence	Mathematical Operation
First	Parentheses. For nested parentheses, the innermost are executed first.
Second	Exponentiation.
Third	Multiplication, division (equal precedence).
Fourth	Addition and subtraction.

In an expression that has several operations, higher-precedence operations are executed before lower-precedence operations. If two or more operations have the same precedence, the expression is executed from left to right. As illustrated in the next section, parentheses can be used to change the order of calculations.

1.3.2 Using MATLAB as a Calculator

The simplest way to use MATLAB is as a calculator. This is done in the Command Window by typing a mathematical expression and pressing the **Enter** key. MATLAB calculates the expression and responds by displaying ans = and the numerical result of the expression in the next line. This is demonstrated in Tutorial 1-1.

Tutorial 1-1: Using MATLAB as a calculator.

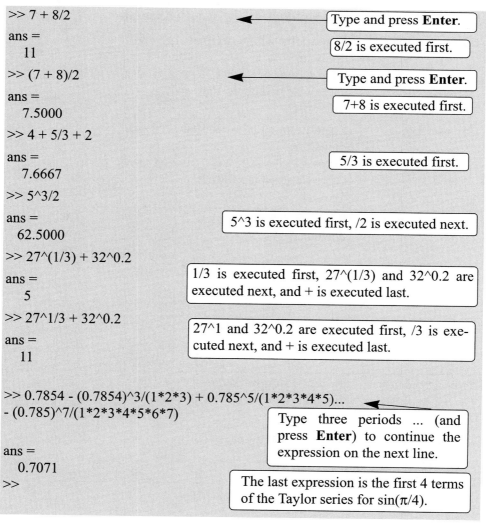

```
>> 7 + 8/2                          ◄———— Type and press Enter.

ans =                                     8/2 is executed first.
    11
>> (7 + 8)/2                         ◄———— Type and press Enter.

ans =                                     7+8 is executed first.
    7.5000
>> 4 + 5/3 + 2

ans =                                     5/3 is executed first.
    7.6667
>> 5^3/2

ans =                               5^3 is executed first, /2 is executed next.
    62.5000
>> 27^(1/3) + 32^0.2

ans =                               1/3 is executed first, 27^(1/3) and 32^0.2 are
    5                               executed next, and + is executed last.

>> 27^1/3 + 32^0.2

ans =                               27^1 and 32^0.2 are executed first, /3 is exe-
    11                              cuted next, and + is executed last.


>> 0.7854 - (0.7854)^3/(1*2*3) + 0.785^5/(1*2*3*4*5)...
 - (0.785)^7/(1*2*3*4*5*6*7)       ◄————
                                    Type three periods ... (and
                                    press Enter) to continue the
ans =                               expression on the next line.
    0.7071
>>                                  The last expression is the first 4 terms
                                    of the Taylor series for sin(π/4).
```

1.4 DISPLAY FORMATS

The user can control the format in which MATLAB displays output on the screen. In Tutorial 1-1, the output format is fixed-point with 4 decimal digits (called short), which is the default format for numerical values. The format can be changed with the format command. Once the format command is entered, all the output that follows is displayed in the specified format. Several of the available formats are listed and described in Table 1-2.

MATLAB has several other formats for displaying numbers. Details of these formats can be obtained by typing help format in the Command Window. The format in which numbers are displayed does not affect how MATLAB computes and saves numbers.

Table 1-2: Display formats

Command	Description	Example
format short	Fixed-point with 4 decimal digits for: $0.001 \leq number \leq 1000$ Otherwise display format short e.	>> 290/7 ans = 41.4286
format long	Fixed-point with 14 decimal digits for: $0.001 \leq number \leq 100$ Otherwise display format long e.	>> 290/7 ans = 41.42857142857143
format short e	Scientific notation with 4 decimal digits.	>> 290/7 ans = 4.1429e+001
format long e	Scientific notation with 15 decimal digits.	>> 290/7 ans = 4.142857142857143e+001
format short g	Best of 5-digit fixed or floating point.	>> 290/7 ans = 41.429
format long g	Best of 15-digit fixed or floating point.	>> 290/7 ans = 41.4285714285714
format bank	Two decimal digits.	>> 290/7 ans = 41.43

Table 1-2: Display formats (Continued)

Command	Description	Example
`format compact`	Eliminates empty lines to allow more lines with information displayed on the screen.	
`format loose`	Adds empty lines (opposite of `compact`).	

1.5 ELEMENTARY MATH BUILT-IN FUNCTIONS

In addition to basic arithmetic operations, expressions in MATLAB can include functions. MATLAB has a very large library of built-in functions. A function has a name and an argument in parentheses. For example, the function that calculates the square root of a number is `sqrt(x)`. Its name is `sqrt`, and the argument is `x`. When the function is used, the argument can be a number, a variable that has been assigned a numerical value (explained in Section 1.6), or a computable expression that can be made up of numbers and/or variables. Functions can also be included in arguments, as well as in expressions. Tutorial 1-2 shows examples of using the function `sqrt(x)` when MATLAB is used as a calculator with scalars.

Tutorial 1-2: Using the `sqrt` built-in function.

```
>> sqrt(64)                          Argument is a number.
ans =
     8
>> sqrt(50 + 14*3)                    Argument is an expression.
ans =
     9.5917
>> sqrt(54 + 9*sqrt(100))             Argument includes a function.
ans =
    12
>> (15 + 600/4)/sqrt(121)            Function is included in an expression.
ans =
    15
>>
```

Lists of some commonly used elementary MATLAB mathematical built-in functions are given in Tables 1-3 through 1-5. A complete list of functions organized by name of category can be found in the Help Window.

Table 1-3: Elementary math functions

Function	Description	Example
sqrt(x)	Square root.	>> sqrt(81) ans = 9
exp(x)	Exponential (e^x).	>> exp(5) ans = 148.4132
abs(x)	Absolute value.	>> abs(-24) ans = 24
log(x)	Natural logarithm. Base e logarithm (ln).	>> log(1000) ans = 6.9078
log10(x)	Base 10 logarithm.	>> log10(1000) ans = 3.0000
factorial(x)	The factorial function $x!$ (x must be a positive integer.)	>> factorial(5) ans = 120

Table 1-4: Trigonometric math functions

Function	Description	Example
sin(x)	Sine of angle x (x in radians).	>> sin(pi/6) ans = 0.5000
cos(x)	Cosine of angle x (x in radians).	>> cos(pi/6) ans = 0.8660
tan(x)	Tangent of angle x (x in radians).	>> tan(pi/6) ans = 0.5774
cot(x)	Cotangent of angle x (x in radians).	>> cot(pi/6) ans = 1.7321

The inverse trigonometric functions are asin(x), acos(x), atan(x), and acot(x). The hyperbolic trigonometric functions are sinh(x), cosh(x), tanh(x), and coth(x). The previous table uses pi which is equal to π (see Section 1.6.3).

Table 1-5: Rounding functions

Function	Description	Example
round(x)	Round to the nearest integer.	>> round(17/5) ans = 3
fix(x)	Round towards zero.	>> fix(13/5) ans = 2
ceil(x)	Round towards infinity.	>> ceil(11/5) ans = 3
floor(x)	Round towards minus infinity.	>> floor(-9/4) ans = -3
rem(x,y)	Returns the remainder after x is divided by y.	>> rem(13,5) ans = 3
sign(x)	Signum function. Returns 1 if $x > 0$, -1 if $x < 0$, and 0 if $x = 0$.	>> sign(5) ans = 1

1.6 DEFINING SCALAR VARIABLES

A variable is a name made of a letter or a combination of several letters (and digits) that is assigned a numerical value. Once a variable is assigned a numerical value, it can be used in mathematical expressions, in functions, and in any MATLAB statements and commands. A variable is actually a name of a memory location. When a new variable is defined, MATLAB allocates an appropriate memory space where the variable's assignment is stored. When the variable is used the stored data is used. If the variable is assigned a new value the content of the memory location is replaced. (In Chapter 1 we only consider variables that are assigned numerical values that are scalars. Assigning and addressing variables that are arrays is discussed in Chapter 2.)

1.6.1 The Assignment Operator

In MATLAB the = sign is called the assignment operator. The assignment operator assigns a value to a variable.

> Variable_name = A numerical value, or a computable expression

- The left-hand side of the assignment operator can include only one variable name.

The right-hand side can be a number, or a computable expression that can include numbers and/or variables that were previously assigned numerical values. When the **Enter** key is pressed the numerical value of the right-hand side is assigned to the variable, and MATLAB displays the variable and its assigned value in the next two lines.

The following shows how the assignment operator works:

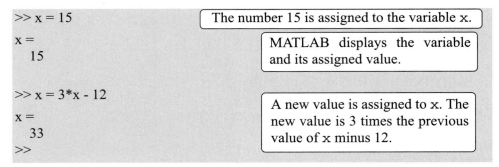

The last statement ($x = 3x - 12$) illustrates the difference between the assignment operator and the equal sign. If in this statement the = sign meant equal, the value of x would be 6 (solving the equation for x).

The use of previously defined variables to define a new variable is demonstrated next.

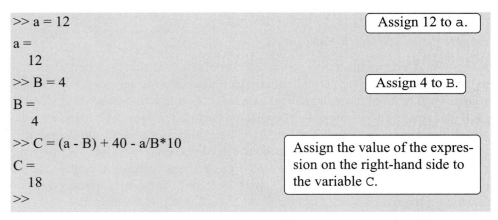

- If a semicolon is typed at the end of the command then, when the **Enter** key is pressed, MATLAB does not display the variable with its assigned value (the variable still exists and is stored in memory).

- If a variable already exists, typing the variable's name and pressing the **Enter** key will display the variable and its value in the next two lines.

For example, the last demonstration is repeated below using semicolons:

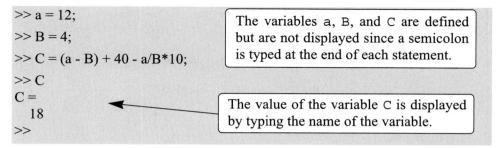

```
>> a = 12;
>> B = 4;
>> C = (a - B) + 40 - a/B*10;
>> C
C =
    18
>>
```

The variables a, B, and C are defined but are not displayed since a semicolon is typed at the end of each statement.

The value of the variable C is displayed by typing the name of the variable.

- Several assignments can be typed in the same line. The assignments must be separated with a comma (spaces can be added after the comma). When the **Enter** key is pressed, the assignments are executed from left to right and the variables and their assignments are displayed. A variable is not displayed if a semicolon is typed instead of a comma. For example, the assignments of the variables a, B, and C above can all be done in the same line.

```
>> a = 12, B = 4; C = (a - B) + 40 - a/B*10
a =
    12
C =
    18
```

The variable B is not displayed because a semi-colon is typed at the end of the assignment.

- A variable that already exists can be reassigned a new value. For example:

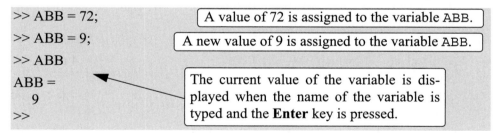

```
>> ABB = 72;
>> ABB = 9;
>> ABB
ABB =
     9
>>
```

A value of 72 is assigned to the variable ABB.

A new value of 9 is assigned to the variable ABB.

The current value of the variable is displayed when the name of the variable is typed and the **Enter** key is pressed.

- Once a variable is defined it can be used as an argument in functions. For example:

```
>> x = 0.75;
>> E = sin(x)^2 + cos(x)^2
E =
     1
>>
```

1.6.2 Rules About Variable Names

- Variable names:

- Can be up to 63 (in MATLAB 7) characters long (31 characters in MATLAB 6.0).

- Can contain letters, digits, and the underscore character.

- Must begin with a letter.

- MATLAB is case sensitive; it distinguishes between uppercase and lowercase letters. For example, AA, Aa, aA, and aa are the names of four different variables.

- Avoid using the names of a built-in function for a variable (i.e. avoid using: cos, sin, exp, sqrt, etc.). Once a function name is used to define a variable, the function cannot be used.

1.6.3 Predefined Variables

A number of frequently used variables are already defined when MATLAB is started. Some of the predefined variables are:

ans A variable that has the value of the last expression that was not assigned to a specific variable (see Tutorial 1-1). If the user does not assign the value of an expression to a variable, MATLAB automatically stores the result in ans.

pi The number π.

eps The smallest difference between two numbers. Equal to $2^{(-52)}$, which is approximately 2.2204e–016.

inf Used for infinity.

i Defined as $\sqrt{-1}$, which is: 0 + 1.0000i.

j Same as i.

NaN Stands for Not-a-Number. Used when MATLAB cannot determine a valid numeric value. For example 0/0.

The predefined variables can be redefined to have any other value. The variables pi, eps, and inf, are usually not redefined since they are frequently used in many applications. Other predefined variables like i and j are sometime redefined (commonly in association with loops) when complex numbers are not involved in the application.

1.7 USEFUL COMMANDS FOR MANAGING VARIABLES

The following are commands that can be used to eliminate variables or to obtain information about variables that have been created. When these commands are typed in the Command Window and the **Enter** key is pressed, they either provide information, or they perform a task as listed below.

Command	Outcome
`clear`	Removes all variables from the memory.
`clear x y z`	Removes only variables x, y, and z from the memory.
`who`	Displays a list of the variables currently in the memory.
`whos`	Displays a list of the variables currently in the memory and their size together with information about their bytes and class.

1.8 EXAMPLES OF MATLAB APPLICATIONS

Sample Problem 1-1:   Trigonometric identity

A trigonometric identity is given by:

$$\cos^2\frac{x}{2} = \frac{\tan x + \sin x}{2\tan x}$$

Verify that the identity is correct by calculating each side of the equation, substituting $x = \frac{\pi}{5}$.

Solution

```
>> x = pi/5;                                      ┌─────────────────────────────┐
                                                  │ Define x.                   │
>> LHS = cos(x/2)^2                               └─────────────────────────────┘
                                       ┌─────────────────────────────────────┐
LHS =                                  │ Calculate the left-hand side.        │
    0.9045                             └─────────────────────────────────────┘
>> RHS = (tan(x) + sin(x))/(2*tan(x))
                                       ┌─────────────────────────────────────┐
RHS =                                  │ Calculate the right-hand side.       │
    0.9045                             └─────────────────────────────────────┘
>>
```

Sample Problem 1-2:   Geometry and trigonometry

Four circles are placed, as shown in the figure. At each point that two circles are in contact they are tangent to each other. Determine the distance between the centers C_2 and C_4.

The radii of the circles are:

R_1 = 16 mm, R_2 = 6.5 mm, R_3 = 12 mm, and R_4 = 9.5 mm.

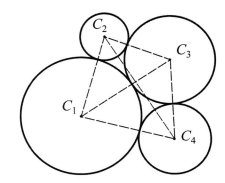

Solution

The lines that connect the centers of the circles create four triangles. In two of the triangles, $\Delta C_1 C_2 C_3$ and $\Delta C_1 C_3 C_4$, the lengths of all the sides are known. This information is used to calculate the angles γ_1 and γ_2 in these triangles by using the law of cosines. For example, γ_1 is calculated from:

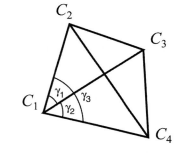

$$(C_2 C_3)^2 = (C_1 C_2)^2 + (C_1 C_3)^2 - 2(C_1 C_2)(C_1 C_3)\cos\gamma_1$$

Next, the length of the side $C_2 C_4$ is calculated by considering the triangle $\Delta C_1 C_2 C_4$. This is done, again, by using the law of cosines (the lengths $C_1 C_2$ and $C_1 C_4$ are known and the angle γ_3 is the sum of the angles γ_1 and γ_2).

```
>> R1 = 16; R2 = 6.5; R3 = 12; R4 = 9.5;                    Define the R's.

>> C1C2 = R1 + R2; C1C3 = R1 + R3; C1C4 = R1 + R4;   Calculate the
>> C2C3 = R2 + R3; C3C4 = R3 + R4;                   lengths of the sides.

>> Gama1 = acos((C1C2^2 + C1C3^2 - C2C3^2)/(2*C1C2*C1C3));

>> Gama2 = acos((C1C3^2 + C1C4^2 - C3C4^2)/(2*C1C3*C1C4));

>> Gama3 = Gama1 + Gama2;                    Calculate γ1, γ2, and γ3.


>> C2C4 = sqrt(C1C2^2 + C1C4^2 - 2*C1C2*C1C4*cos(Gama3))

C2C4 =                                       Calculate the length of
   33.5051                                   side C2C4.
```

Sample Problem 1-3:   Heat transfer

An object with an initial temperature of T_0 that is placed at time $t = 0$ inside a chamber that has a constant temperature of T_s, will experience a temperature change according to the equation:

$$T = T_s + (T_0 - T_s)e^{-kt}$$

where T is the temperature of the object at time t, and k is a constant. A soda can at a temperature of 120°F (was left in the car) is placed inside a refrigerator where the temperature is 38°F. Determine, to the nearest degree, the temperature of the can after three hours. Assume $k = 0.45$. First define all the variables and then calculate the temperature using one MATLAB command.

Solution

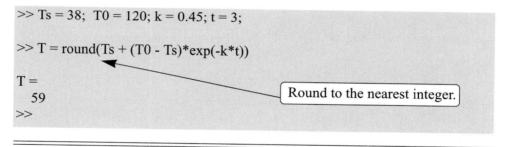

```
>> Ts = 38;  T0 = 120; k = 0.45; t = 3;

>> T = round(Ts + (T0 - Ts)*exp(-k*t))

T =
   59
>>
```

Round to the nearest integer.

Sample Problem 1-4:   Compounded interest

The balance B of a savings account after t years when a principal P is invested at an annual interest rate r and the interest is compounded n times a year is given by:

$$B = P\left(1 + \frac{r}{n}\right)^{nt} \qquad (1)$$

If the interest is compounded yearly, the balance is given by:

$$B = P(1 + r)^t \qquad (2)$$

In one account $5,000 is invested for 17 years in an account where the interest is compounded yearly. In a second account $5,000 is invested in an account in which the interest is compounded monthly. In both accounts the interest rate is 8.5%. Use MATLAB to determine how long (in years and months) it would take for the balance in the second account to be the same as the balance of the first account after 17 years.

Solution

Follow these steps:
(a) Calculate B for $5,000 invested in a yearly compounded interest account after 17 years using Equation (2).
(b) Calculate the t for the B calculated in part (a), from the monthly compounded

interest formula, Equation (1).

(*c*) Determine the number of years and months that correspond to *t*.

```
>> P = 5000;  r = 0.085;  ta = 17; n = 12;
>> B = P*(1 + r)^ta                      Step (a): Calculate B from Eq. (2).
B =
   2.0011e+004
>> t = log(B/P)/(n*log(1 + r/n))         Step (b): Solve Eq. (1)
t =                                      for t, and calculate t.
   16.3737
>> years = fix(t)                        Step (c): Determine the number of years.
years =
   16
>> months = ceil((t - years)*12)         Determine the number of months.
months =
   5
```

1.9 PROBLEMS

Solve the following problems in the Command Window.

1. Calculate:

    a) $\dfrac{35.7 \cdot 64 - 7^3}{45 + 5^2}$

    b) $\dfrac{5}{4} \cdot 7 \cdot 6^2 + \dfrac{3^7}{(9^3 - 652)}$

2. Calculate:

    a) $(2 + 7)^3 + \dfrac{273^{2/3}}{2} + \dfrac{55^2}{3}$

    b) $2^3 + 7^3 + \dfrac{273^3}{2} + 55^{3/2}$

3. Calculate:

    a) $\dfrac{3^7 \log(76)}{7^3 + 546} + \sqrt[3]{910}$

b) $43 \cdot \dfrac{\left(\sqrt[4]{250}+23\right)^2}{e^{\left(45-3^3\right)}}$

4.   Calculate:

a) $\cos^2\left(\dfrac{5\pi}{6}\right)\sin\left(\dfrac{7\pi}{8}\right)^2 + \dfrac{\tan\left(\dfrac{\pi}{6}\ln 8\right)}{\sqrt{7}}$

b) $\cos\left(\dfrac{5\pi}{6}\right)^2\sin^2\left(\dfrac{7\pi}{8}\right) + \dfrac{\tan\left(\dfrac{\pi\ln 8}{6}\right)}{7\cdot\dfrac{5}{2}}$

5.   Define the variable x as $x = 13.5$, then evaluate:

a) $x^3 + 5x^2 - 26.7x - 52$

b) $\dfrac{\sqrt{14x^3}}{e^{3x}}$

c) $\log\left|x^2 - x^3\right|$

6.   Define the variables x and z as $x = 9.6$, and $z = 8.1$, then evaluate:

a) $xz^2 - \left(\dfrac{2z}{3x}\right)^{\frac{3}{5}}$

b) $\dfrac{443z}{2x^3} + \dfrac{e^{-xz}}{(x+z)}$

7.   Define the variables a, b, c, and d as:
     $a = 15.62$, $b = -7.08$, $c = 62.5$ and $d = 0.5(ab - c)$.
     Evaluate:

a) $a + \dfrac{ab(a+d)^2}{c}\dfrac{}{\sqrt{|ab|}}$

b) $de^{\left(\frac{d}{2}\right)} + \dfrac{\dfrac{ad+cd}{20} + \dfrac{30}{b}}{\dfrac{a}{(a+b+c+d)}}$

8.   Calculate (by writing one command) the radius r of a sphere that has a volume of 350 in³. Once r is determined, use it to calculate the surface area of the sphere.

9. Two trigonometric identities are given by:

    a) $\sin 2x = 2\sin x \cos x$

    b) $\cos\dfrac{x}{2} = \sqrt{\dfrac{1 + \cos x}{2}}$

    For each part, verify that the identity is correct by calculating each side of the equation, substituting $x = \dfrac{5}{24}\pi$.

10. Two trigonometric identities are given by:

    a) $\tan 2x = \dfrac{2\tan x}{1 - \tan^2 x}$

    b) $\tan\dfrac{x}{2} = \sqrt{\dfrac{1 - \cos x}{1 + \cos x}}$

    For each part, verify that the identity is correct by calculating the values of the left and right sides of the equation, substituting $x = \dfrac{3}{17}\pi$.

11. Define two variables: *alpha* = 5π/9, *beta* = π/7. Using these variables, show that the following trigonometric identity is correct by calculating the value of the left and right sides of the equation.

    $$\cos\alpha - \cos\beta = 2\sin\frac{1}{2}(\alpha + \beta)\sin\frac{1}{2}(\beta - \alpha)$$

12. In the right triangle shown $a = 11$ cm, and $c = 21$ cm. Define a and c as variables, and then:

    a) Using the Pythagorean Theorem, calculate b by typing one line in the Command Window.

    b) Using b from part a), and the `acos(x)` function, calculate the angle α in degrees, typing one line in the Command Window.

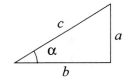

13. In the triangle shown $a = 18$ cm, $b = 35$ cm, and $c = 50$ cm. Define a, b, and c as variables, and then calculate the angle γ (in degrees) by substituting the variables in the Law of Cosines.

    (The Law of Cosines: $c^2 = a^2 + b^2 - 2ab\cos\gamma$)

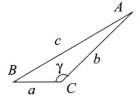

14. The distance d from a point (x_0, y_0) to a line $Ax + By + C = 0$ is given by:

$$d = \frac{|Ax_0 + By_0 + C|}{\sqrt{A^2 + B^2}}$$

Determine the distance of the point $(2, -3)$ from the line $3x + 5y - 6 = 0$. First define the variables A, B, C, x_0, and y_0, and then calculate d. (Use the abs and sqrt functions.)

15. Flowers are packed in boxes such that a dozen are placed in each box. Determine how many boxes are needed to pack 751 flowers, using the ceil function.

16. Define the following variables:

    table_price = $256.95
    chair_price = $89.99

Then change the display format to bank and:

    a) Evaluate the cost of two tables and eight chairs.
    b) The same as part a), but add 5.5% sale tax.
    c) The same as part b) but round the total cost to the nearest dollar.

17. When adding fractions, the smallest common denominator must be determined. For example, the smallest common denominator of 1/4 and 1/10 is 20. Use the MATLAB Help Window to find a MATLAB built-in function that determines the least common multiplier of two numbers. Then use the function to show that the least common multiplier of:

    a) 4 and 10 is 20.
    b) 6 and 38 is 114.

18. The magnitude M of an earthquake on the Richter scale is given by: $M = \frac{2}{3}\log\frac{E}{E_0}$, where E is the energy released by the earthquake, and $E_0 = 10^{4.4}$ Joules is a constant (energy of a small reference earthquake). Determine how many times more energy is released from an earthquake that registers 7.2 on the Richter scale than an earthquake that registers 5.3.

Chapter 2
Creating Arrays

The array is a fundamental form that MATLAB uses to store and manipulate data. An array is a list of numbers arranged in rows and/or columns. The simplest array (one-dimensional) is a row, or a column of numbers. A more complex array (two-dimensional) is a collection of numbers arranged in rows and columns. One use of arrays is to store information and data, as in a table. In science and engineering, one-dimensional arrays frequently represent vectors, and two-dimensional arrays often represent matrices. Chapter 2 shows how to create and address arrays while Chapter 3 shows how to use arrays in mathematical operations. In addition to arrays that are made of numbers, arrays in MATLAB can also be made of a list of characters, which are called strings. Strings are discussed in Section 2.10.

2.1 CREATING A ONE-DIMENSIONAL ARRAY (VECTOR)

A one-dimensional array is a list of numbers that is placed in a row or a column. One example is the representation of the position of a point in space in a three-dimensional Cartesian coordinate system. As shown in Figure 2-1, the position of point A is defined by a list of three numbers 2, 4, and 5, which are the coordinates of the point.

The position of point A can be expressed in terms of a position vector:

$$\mathbf{r}_A = 2\mathbf{i} + 4\mathbf{j} + 5\mathbf{k}$$

where $\mathbf{i}$, $\mathbf{j}$, and $\mathbf{k}$ are unit vectors in the direction of the x, y, and z axis, respectively. The numbers 2, 4, and 5 can be used to define a row or a column vector.

Any list of numbers can be set up as a vector. For example, Table 2-1 contains population growth data that can be used to create two lists of numbers; one of the years and the other of the popula-

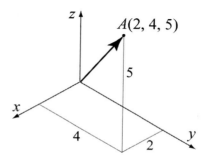

Figure 2-1: Position of a point.

tion. Each list can be entered as elements in a vector with the numbers placed in a row or in a column.

Table 2-1: Population data

Year	1984	1986	1988	1990	1992	1994	1996
Population (Millions)	127	130	136	145	158	178	211

In MATLAB, a vector is created by assigning the elements of the vector to a variable. This can be done in several ways depending on the source of the information that is used for the elements of the vector. When a vector contains specific numbers that are known (like the coordinates of point *A*), the value of each element is entered directly. Each element can also be a mathematical expression that can include predefined variables, numbers, and functions. Often, the elements of a row vector are a series of numbers with constant spacing. In such cases the vector can be created with MATLAB commands. A vector can also be created as the result of mathematical operations as explained in Chapter 3.

Creating a vector from a known list of numbers:

The vector is created by typing the elements (numbers) inside square brackets [ ].

```
variable_name = [ type vector elements ]
```

Row vector: To create a row vector type the elements with a space or a comma between the elements inside the square brackets.

Column vector: To create a column vector type the left square bracket [and then enter the elements with a semicolon between them, or press the **Enter** key after each element. Type the right square bracket] after the last element.

Tutorial 2-1 shows how the data from Table 2-1 and the coordinates of point *A* are used to create row and column vectors.

Tutorial 2-1: Creating vectors from given data.

```
>> yr = [1984 1986 1988 1990 1992 1994 1996]
```
The list of years is assigned to a row vector named yr.
```
yr =
      1984      1986      1988      1990      1992      1994      1996
>> pop = [127;  130;  136;  145;  158;  178;  211]
```
The population data is assigned to a column vector named pop.
```
pop =
   127
   130
   136
   145
   158
```

Tutorial 2-1: Creating vectors from given data. (Continued)

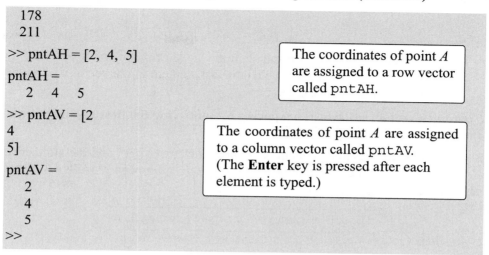

```
    178
    211
>> pntAH = [2, 4, 5]
pntAH =
    2    4    5
>> pntAV = [2
4
5]
pntAV =
    2
    4
    5
>>
```

> The coordinates of point *A* are assigned to a row vector called `pntAH`.

> The coordinates of point *A* are assigned to a column vector called `pntAV`. (The **Enter** key is pressed after each element is typed.)

Creating a vector with constant spacing by specifying the first term, the spacing, and the last term:

In a vector with constant spacing the difference between the elements is the same. For example, in the vector: $v = 2\ 4\ 6\ 8\ 10$, the spacing between the elements is 2. A vector in which the first term is *m*, the spacing is *q*, and the last term is *n* is created by typing:

```
variable_name = [m:q:n]
```
or
```
variable_name = m:q:n
```

(The brackets are optional.)

Some examples are:

```
>> x = [1:2:13]
x =
    1    3    5    7    9    11    13
```
> First element 1, spacing 2, last element 13.

```
>> y = [1.5:0.1:2.1]
y =
    1.5000    1.6000    1.7000    1.8000    1.9000    2.0000    2.1000
```
> First element 1.5, spacing 0.1, last element 2.1.

```
>> z = [-3:7]
z =
    -3    -2    -1    0    1    2    3    4    5    6    7
```
> First element −3, last term 7. If spacing is omitted, the default is 1.

```
>> xa = [21:-3:6]
```
> First element 21, spacing −3, last term 6.

```
xa =
    21   18   15   12   9   6
>>
```

- If the numbers m, q, and n are such that the value of n can not be obtained by adding q's to m, then (for positive n) the last element in the vector will be the last number that does not exceed n.

Creating a vector with constant spacing by specifying the first and last terms, and the number of terms:

A vector in which the first element is *xi*, the last element is *xf*, and the number of elements is *n* is created by typing the linspace command (MATLAB determines the correct spacing):

$$\boxed{\texttt{variable\_name = linspace(xi,xf,n)}}$$

Some examples are:

```
>> va = linspace(0,8,6)          6 elements, first element 0, last element 8.
va =
         0   1.6000   3.2000   4.8000   6.4000   8.0000
>> vb = linspace(30,10,11)       11 elements, first element 30, last element 10.
vb =
    30   28   26   24   22   20   18   16   14   12   10
>> u = linspace(49.5,0.5)        First element 49.5, last element 0.5.

                                 When the number of elements is
u =                              omitted, the default is 100.
 Columns 1 through 10
   49.5000   49.0051   48.5101   48.0152   47.5202   47.0253   46.5303   46.0354
 45.5404   45.0455
 ...........                     100 elements are displayed.
 Columns 91 through 100
    4.9545    4.4596    3.9646    3.4697    2.9747    2.4798    1.9848    1.4899
 0.9949   0.5000
>>
```

2.2 CREATING A TWO-DIMENSIONAL ARRAY (MATRIX)

A two-dimensional array, also called a matrix, has numbers in rows and columns. Matrices can be used to store information like in a table. Matrices play an important role in linear algebra and are used in science and engineering to describe many physical quantities.

In a square matrix the number of rows and columns is equal. For example, the matrix:

7 4 9
3 8 1 3×3 matrix
6 5 3

is square, with three rows and three columns. In general, the number of rows and columns can be different. For example, the matrix:

31 26 14 18 5 30
 3 51 20 11 43 65 4×6 matrix
28 6 15 61 34 22
14 58 6 36 93 7

has four rows and six columns. A $m \times n$ matrix has m rows and n columns, and m by n is called the size of the matrix.

A matrix is created by assigning the elements of the matrix to a variable. This is done by typing the elements, row by row, inside square brackets [ ]. First type the left bracket [, then type the first row separating the elements with spaces or commas. To type the next row type a semicolon or press **Enter**. Type the right bracket] at the end of the last row.

> variable_name = [1st row elements; 2nd row elements; 3rd row elements; ..... ; last row elements]

The elements that are entered can be numbers or mathematical expressions that may include numbers, predefined variables, and functions. All the rows must have the same number of elements. If an element is zero, it has to be entered as such. MATLAB displays an error message if an attempt is made to define an incomplete matrix. Examples of matrices defined in different ways are shown in Tutorial 2-2.

Tutorial 2-2: Creating matrices.

```
>> a = [5 35 43; 4 76 81; 21 32 40]
a =
    5   35   43
    4   76   81
   21   32   40
>> b = [7 2 76 33 8
1 98 6 25 6
5 54 68 9 0]
b =
    7    2   76   33    8
    1   98    6   25    6
    5   54   68    9    0
```

A semicolon is typed before a new line is entered.

The **Enter** key is pressed before a new line is entered.

Tutorial 2-2: Creating matrices. (Continued)

```
>> cd = 6; e = 3; h = 4;                          Three variables are defined.
>> Mat = [e, cd*h, cos(pi/3); h^2, sqrt(h*h/cd), 14]
Mat =
    3.0000   24.0000    0.5000                    Elements are defined
   16.0000    1.6330   14.0000                    by mathematical
>>                                                 expressions.
```

Rows of a matrix can also be entered as vectors using the notation for creating vectors with constant spacing, or the `linspace` command. For example:

```
>> A=[1:2:11; 0:5:25; linspace(10,60,6); 67 2 43 68 4 13]
A =
     1     3     5     7     9    11
     0     5    10    15    20    25
    10    20    30    40    50    60
    67     2    43    68     4    13
>>
```

In this example the first two rows were entered as vectors using the notation of constant spacing, the third row was entered using the `linspace` command, and in the last row the elements were entered individually.

2.2.1 The `zeros`, `ones` and `eye` Commands

The `zeros(m,n)`, the `ones(m,n)`, and `eye(n)` commands can be used to create matrices that have elements with special values. The `zeros(m,n)` and the `ones(m,n)` commands create a matrix with *m* rows and *n* columns, in which all the elements are the numbers 0 and 1, respectively. The `eye(n)` command creates a square matrix with *n* rows and *n* columns in which the diagonal elements are equal to 1, and the rest of the elements are 0. This matrix is called the identity matrix. Examples are:

```
>> zr = zeros(3,4)
zr =
     0     0     0     0
     0     0     0     0
     0     0     0     0
>> ne = ones(4,3)
ne =
     1     1     1
     1     1     1
     1     1     1
     1     1     1
```

```
>> idn = eye(5)
idn =
     1    0    0    0    0
     0    1    0    0    0
     0    0    1    0    0
     0    0    0    1    0
     0    0    0    0    1
>>
```

Matrices can also be created as a result of mathematical operations with vectors and matrices. This topic is covered in Chapter 3.

2.3 NOTES ABOUT VARIABLES IN MATLAB

- All variables in MATLAB are arrays. A scalar is an array with one element, a vector is an array with one row, or one column, of elements, and a matrix is an array with elements in rows and columns.

- The variable (scalar, vector, or matrix) is defined by the input when the variable is assigned. There is no need to define the size of the array (single element for a scalar, a row or a column of elements for a vector, or a two-dimensional array of elements for a matrix) before the elements are assigned.

- Once a variable exists, as a scalar, vector, or a matrix, it can be changed to be any other size, or type, of variable. For example, a scalar can be changed to be a vector or a matrix, a vector can be changed to be a scalar, a vector of different length, or a matrix, and a matrix can be changed to have a different size, or to be reduced to a vector or a scalar. These changes are made by adding or deleting elements. This subject is covered in Sections 2.7 and 2.8.

2.4 THE TRANSPOSE OPERATOR

The transpose operator, when applied to a vector, switches a row (column) vector to a column (row) vector. When applied to a matrix, it switches the rows (columns) to columns (rows). The transpose operator is applied by typing a single quote ′ following the variable to be transposed. Examples are:

```
>> aa = [3  8  1]
aa =
     3    8    1
>> bb = aa'
bb =
     3
     8
     1
```

| Define a row vector aa. |

| Define a column vector bb as the transpose of vector aa. |

```
>> C = [2 55 14 8; 21 5 32 11; 41 64 9 1]
C =
    2   55   14    8
   21    5   32   11
   41   64    9    1
>> D = C'
D =
    2   21   41
   55    5   64
   14   32    9
    8   11    1
>>
```

Define a matrix C with 3 rows and 4 columns.

Define a matrix D as the transpose of matrix C. (D has 4 rows and 3 columns.)

2.5 ARRAY ADDRESSING

Elements in an array (either vector or matrix) can be addressed individually or in subgroups. This is useful when there is a need to redefine only some of the elements, to use specific elements in calculations, or when a subgroup of the elements is used to define a new variable.

2.5.1 Vector

The address of an element in a vector is its position in the row (or column). For a vector named ve, ve (k) refers to the element in position k. The first position is 1. For example, if the vector ve has nine elements:

$ve = 35\ 46\ 78\ 23\ 5\ 14\ 81\ 3\ 55$

then

$ve(4) = 23$, $ve(7) = 81$, and $ve(1) = 35$.

A single vector element, $v(k)$, can be used just as a variable. For example, it is possible to change the value of only one element of a vector by reassigning a new value to a specific address. This is done by typing: $v(k) = value$. A single element can also be used as a variable in a mathematical expression. Examples are:

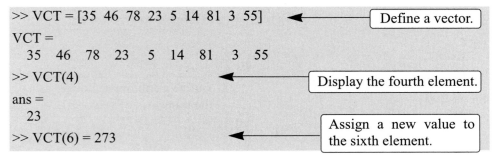

```
>> VCT = [35 46 78 23 5 14 81 3 55]
VCT =
   35   46   78   23    5   14   81    3   55
>> VCT(4)
ans =
   23
>> VCT(6) = 273
```

Define a vector.

Display the fourth element.

Assign a new value to the sixth element.

```
VCT =
    35   46   78   23   5  273  81   3   55        The whole vector is displayed.
>> VCT(2) + VCT(8)

ans =
    49                                             Use the vector elements in
>> VCT(5)^VCT(8) + sqrt(VCT(7))                    mathematical expressions.

ans =
   134
>>
```

2.5.2 Matrix

The address of an element in a matrix is its position, defined by the row number and the column number where it is located. For a matrix assigned to a variable *ma*, $ma(k,p)$ refers to the element in row k and column p.

For example, if the matrix is:
$$ma = \begin{bmatrix} 3 & 11 & 6 & 5 \\ 4 & 7 & 10 & 2 \\ 13 & 9 & 0 & 8 \end{bmatrix}$$

then, $ma(1,1) = 3$, and $ma(2,3) = 10$.

As with vectors, it is possible to change the value of just one element of a matrix by assigning a new value to that element. Also, single elements can be used like variables in mathematical expressions and functions. Some examples are:

```
>> MAT = [3 11 6 5; 4 7 10 2; 13 9 0 8]      Create a 3 × 4 matrix.
MAT =
    3   11    6    5
    4    7   10    2
   13    9    0    8
>> MAT(3,1) = 20                             Assign a new value to the (3,1) element.
MAT =
    3   11    6    5
    4    7   10    2
   20    9    0    8
>> MAT(2,4) - MAT(1,2)                        Use elements in a mathematical expression.

ans =
   -9
>>
```

2.6 USING A COLON : IN ADDRESSING ARRAYS

A colon can be used to address a range of elements in a vector or a matrix.

For a vector:

va(:)   Refers to all the elements of the vector *va* (either a row or a column vector).

va(*m:n*)   Refers to elements *m* through *n* of the vector *va*.

Example:

```
>> v = [4 15 8 12 34 2 50 23 11]          A vector v is created.
v =
   4   15   8   12   34   2   50   23   11
>> u = v(3:7)                    A vector u is created from the ele-
u =                             ments 3 through 7 of vector v.
   8   12   34   2   50
>>
```

For a matrix:

A(:,*n*)   Refers to the elements in all the rows of column *n* of the matrix *A*.

A(*n*,:)   Refers to the elements in all the columns of row *n* of the matrix *A*.

A(:,*m:n*)   Refers to the elements in all the rows between columns *m* and *n* of the matrix *A*.

A(*m:n*,:)   Refers to the elements in all the columns between rows *m* and *n* of the matrix *A*.

A(*m:n,p:q*)   Refers to the elements in rows *m* through *n* and columns *p* through *q* of the matrix *A*.

The use of the colon symbol in addressing elements of matrices is demonstrated in Tutorial 2-3.

Tutorial 2-3: Using a colon in addressing arrays.

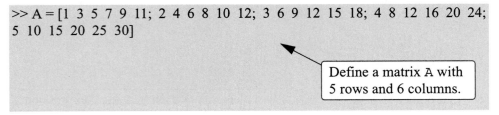

```
>> A = [1 3 5 7 9 11; 2 4 6 8 10 12; 3 6 9 12 15 18; 4 8 12 16 20 24;
5 10 15 20 25 30]
```

Define a matrix A with 5 rows and 6 columns.

Tutorial 2-3: Using a colon in addressing arrays. (Continued)

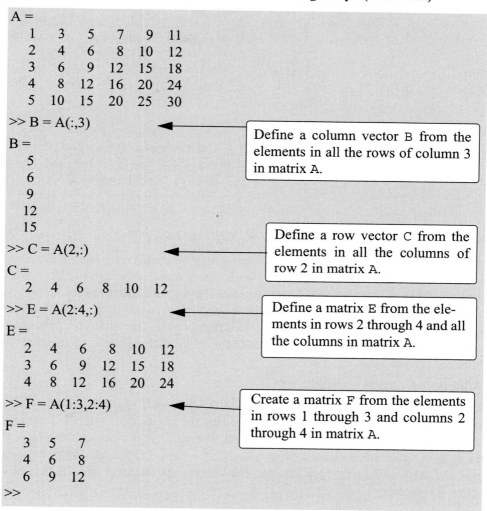

```
A =
    1    3    5    7    9   11
    2    4    6    8   10   12
    3    6    9   12   15   18
    4    8   12   16   20   24
    5   10   15   20   25   30
>> B = A(:,3)

B =
    5
    6
    9
   12
   15
>> C = A(2,:)

C =
    2    4    6    8   10   12
>> E = A(2:4,:)

E =
    2    4    6    8   10   12
    3    6    9   12   15   18
    4    8   12   16   20   24
>> F = A(1:3,2:4)

F =
    3    5    7
    4    6    8
    6    9   12
>>
```

Define a column vector B from the elements in all the rows of column 3 in matrix A.

Define a row vector C from the elements in all the columns of row 2 in matrix A.

Define a matrix E from the elements in rows 2 through 4 and all the columns in matrix A.

Create a matrix F from the elements in rows 1 through 3 and columns 2 through 4 in matrix A.

In Tutorial 2-3 new vectors and matrices were created from existing ones by using a range of elements, or a range of rows and columns (using :). It is possible, however, to select only specific elements, or specific rows and columns of existing variables to create new variables. This is done by typing the selected elements or rows or columns inside brackets, as shown below:

```
>> v = 4:3:34
                                Create a vector v with 11 elements.
v =
    4    7   10   13   16   19   22   25   28   31   34
>> u = v([3, 5, 7:10])
```

Create a vector u from the 3rd, the 5th, and 7th through 10th elements of v.

```
u =
   10   16   22   25   28   31
>> A = [10:-1:4; ones(1,7); 2:2:14; zeros(1,7)]
A =
   10    9    8    7    6    5    4
    1    1    1    1    1    1    1
    2    4    6    8   10   12   14
    0    0    0    0    0    0    0
>> B = A([1,3],[1,3,5:7])
B =
   10    8    6    5    4
    2    6   10   12   14
```

Create a 4×7 matrix A.

Create a matrix B from the 1st and 3rd rows, and 1st, 3rd, and 5th through 7th columns of A.

2.7 ADDING ELEMENTS TO EXISTING VARIABLES

A variable that exists as a vector, or a matrix, can be changed by adding elements to it (remember that a scalar is a vector with one element). A vector (a matrix with a single row or column) can be changed to have more elements, or it can be changed to be a two-dimensional matrix. Rows and/or columns can also be added to an existing matrix to obtain a matrix of different size. The addition of elements can be done by simply assigning values to the additional elements, or by appending existing variables.

Adding elements to a vector:

Elements can be added to an existing vector by assigning values to the new elements. For example, if a vector has 4 elements, the vector can be made longer by assigning values to elements 5, 6, and so on. If a vector has n elements and a new value is assigned to an element with address of $n + 2$ or larger, MATLAB assigns zeros to the elements that are between the last original element and the new element. Examples:

```
>> DF = 1:4
DF =
    1    2    3    4
>> DF(5:10) = 10:5:35
DF =
    1    2    3    4   10   15   20   25   30   35
>> AD = [5 7 2]
AD =
    5    7    2
>> AD(8) = 4
```

Define vector DF with 4 elements.

Adding 6 elements starting with the 5th.

Define vector AD with 3 elements.

Assign a value to the 8th element.

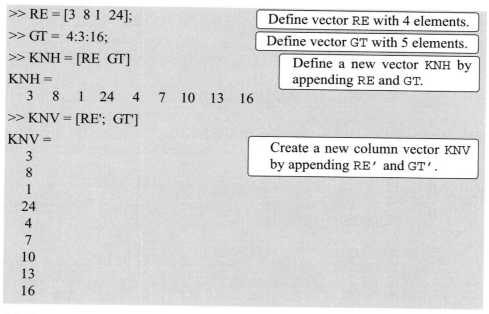

```
AD =
   5   7   2   0   0   0   0   4
>> AR(5) = 24
AR =
   0   0   0   0   24
>>
```

MATLAB assigns zeros to the 4th through 7th elements.

Assign a value to the 5th element of a new vector.

MATLAB assigns zeros to the 1st through 4th elements.

Elements can also be added to a vector by appending existing vectors. Two examples are:

```
>> RE = [3  8  1  24];
>> GT = 4:3:16;
>> KNH = [RE  GT]
KNH =
   3   8   1   24   4   7   10   13   16
>> KNV = [RE';  GT']
KNV =
    3
    8
    1
   24
    4
    7
   10
   13
   16
```

Define vector RE with 4 elements.

Define vector GT with 5 elements.

Define a new vector KNH by appending RE and GT.

Create a new column vector KNV by appending RE′ and GT′.

Adding elements to a matrix:

Rows and/or columns can be added to an existing matrix by assigning values to the new rows or columns. This can be done by assigning new values, or by appending existing variables. This must be done carefully since the size of the added rows or columns must fit the existing matrix. Examples are:

```
>> E = [1 2  3  4;  5  6  7  8]
E =
   1   2   3   4
   5   6   7   8
>> E(3,:) = [10:4:22]
```

Define a 2 × 4 matrix E.

Add the vector 10 14 18 22 as the third row of E.

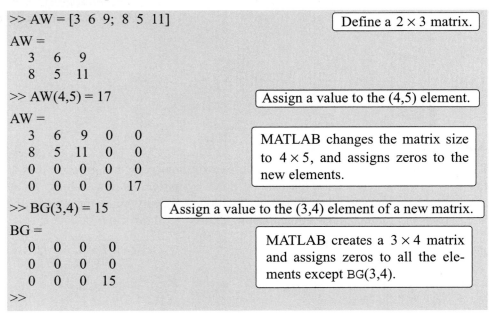

```
E =
    1    2    3    4
    5    6    7    8
   10   14   18   22
>> K = eye(3)                              Define 3 × 3 matrix K.
K =
    1    0    0
    0    1    0
    0    0    1
>> G = [E  K]                              Append the matrix K to
G =                                        matrix E. The number of
                                           rows in E and K must be
    1    2    3    4    1    0    0         the same.
    5    6    7    8    0    1    0
   10   14   18   22    0    0    1
```

If a matrix has a size of $m \times n$, and a new value is assigned to an element with an address beyond the size of the matrix, MATLAB increases the size of the matrix to include the new element. Zeros are assigned to the other elements that are added. Examples:

```
>> AW = [3  6  9;  8  5  11]               Define a 2 × 3 matrix.
AW =
    3    6    9
    8    5   11
>> AW(4,5) = 17                            Assign a value to the (4,5) element.
AW =
    3    6    9    0    0                   MATLAB changes the matrix size
    8    5   11    0    0                   to 4 × 5, and assigns zeros to the
    0    0    0    0    0                   new elements.
    0    0    0    0   17
>> BG(3,4) = 15                            Assign a value to the (3,4) element of a new matrix.
BG =                                       MATLAB creates a 3 × 4 matrix
    0    0    0    0                        and assigns zeros to all the ele-
    0    0    0    0                        ments except BG(3,4).
    0    0    0   15
>>
```

2.8 DELETING ELEMENTS

An element, or a range of elements, of an existing variable can be deleted by reassigning nothing to these elements. This is done by using square brackets with nothing typed in between them. By deleting elements a vector can be made shorter and a matrix can be made to have a smaller size. Examples are:

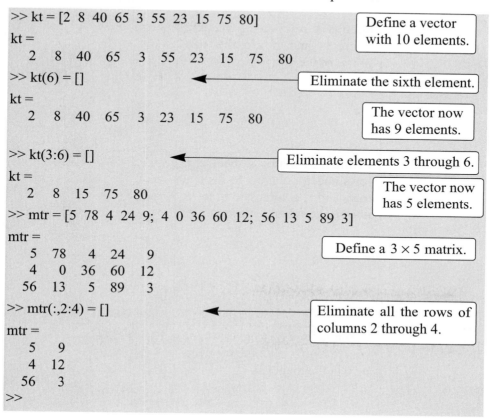

```
>> kt = [2  8  40  65  3  55  23  15  75  80]
kt =
    2   8  40  65   3  55  23  15  75  80
>> kt(6) = []                                          ◄─── Eliminate the sixth element.
kt =
    2   8  40  65   3  23  15  75  80
                                                       The vector now has 9 elements.
>> kt(3:6) = []                                        ◄─── Eliminate elements 3 through 6.
kt =
    2   8  15  75  80
                                                       The vector now has 5 elements.
>> mtr = [5  78  4  24  9;  4  0  36  60  12;  56  13  5  89  3]
mtr =
    5  78   4  24   9
    4   0  36  60  12
   56  13   5  89   3
                                                       Define a 3 × 5 matrix.
>> mtr(:,2:4) = []                                     ◄─── Eliminate all the rows of columns 2 through 4.
mtr =
    5   9
    4  12
   56   3
>>
```

Define a vector with 10 elements.

2.9 BUILT-IN FUNCTIONS FOR HANDLING ARRAYS

MATLAB has many built-in functions for managing and handling arrays. Some of these are listed below:

Table 2-2: Built-in functions for handling arrays

Function	Description	Example
`length(A)`	Returns the number of elements in the vector A.	>> A = [5 9 2 4]; >> length(A) ans = 4

Table 2-2: Built-in functions for handling arrays (Continued)

Function	Description	Example
size(A)	Returns a row vector [m,n], where m and n are the size $m \times n$ of the array A.	>> A = [6 1 4 0 12; 5 19 6 8 2] A = 6 1 4 0 12 5 19 6 8 2 >> size(A) ans = 2 5
reshape(A, m,n)	Rearrange a matrix A that has r rows and s columns to have m rows and n columns. r times s must be equal to m times n.	>> A = [5 1 6; 8 0 2] A = 5 1 6 8 0 2 >> B = reshape(A,3,2) B = 5 0 8 6 1 2
diag(v)	When v is a vector, creates a square matrix with the elements of v in the diagonal.	>> v = [7 4 2]; >> A = diag(v) A = 7 0 0 0 4 0 0 0 2
diag(A)	When A is a matrix, creates a vector from the diagonal elements of A.	>> A = [1 2 3; 4 5 6; 7 8 9] A = 1 2 3 4 5 6 7 8 9 >> vec = diag(A) vec = 1 5 9

Additional built-in functions for manipulation of arrays are described in the Help Window. In this window select "Functions by Category", then "Mathematics", and then "Arrays and Matrices."

Recall from Chapter 1 that the who command displays a list of the variables currently in the memory. The whos command displays a list of the variables currently in the memory and information about their size, bytes, and class. An example is shown below.

```
>> a = 7;
>> E = 3;
>> d = [5, a+E, 4, E^2]
d =
   5   10   4   9
>> g = [a, a^2, 13; a*E, 1, a^E]
g =
    7   49   13
   21    1  343
>> who

Your variables are:
E a d g
>> whos
Name      Size       Bytes  Class
  E        1x1           8  double array
  a        1x1           8  double array
  d        1x4          32  double array
  g        2x3          48  double array

Grand total is 12 elements using 96 bytes
>>
```

> Creating the variables a, E, d, and g.

> The who command displays the variables currently in the memory.

> The whos command displays the variables currently in the memory, and information about their size.

Sample Problem 2-1: Create a matrix

Using the ones and zeros commands, create a 4×5 matrix in which the first two rows are 0's and the next two rows are 1's.

Solution

```
>> A(1:2,:) = zeros(2,5)
A =
   0   0   0   0   0
   0   0   0   0   0
>> A(3:4,:) = ones(2,5)
```

> First, create a 2×5 matrix with zeros.

> Add rows 3 and 4 with ones.

```
A =
    0    0    0    0    0
    0    0    0    0    0
    1    1    1    1    1
    1    1    1    1    1
>>
```

Sample Problem 2-2: Create a matrix

Create a 6×6 matrix in which the middle two rows, and the middle two columns are 1's, and the rest are 0's.

Solution

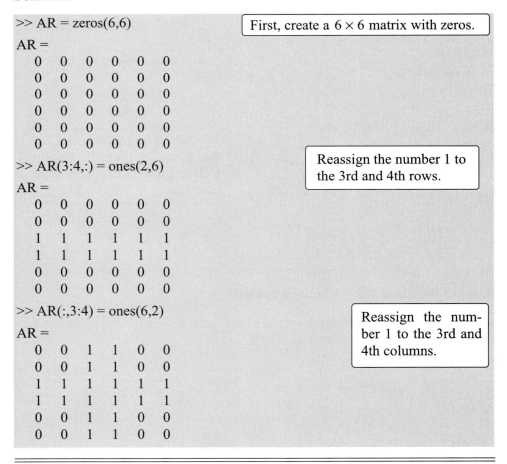

```
>> AR = zeros(6,6)
```
First, create a 6×6 matrix with zeros.

```
AR =
    0    0    0    0    0    0
    0    0    0    0    0    0
    0    0    0    0    0    0
    0    0    0    0    0    0
    0    0    0    0    0    0
    0    0    0    0    0    0
>> AR(3:4,:) = ones(2,6)
```
Reassign the number 1 to the 3rd and 4th rows.

```
AR =
    0    0    0    0    0    0
    0    0    0    0    0    0
    1    1    1    1    1    1
    1    1    1    1    1    1
    0    0    0    0    0    0
    0    0    0    0    0    0
>> AR(:,3:4) = ones(6,2)
```
Reassign the number 1 to the 3rd and 4th columns.

```
AR =
    0    0    1    1    0    0
    0    0    1    1    0    0
    1    1    1    1    1    1
    1    1    1    1    1    1
    0    0    1    1    0    0
    0    0    1    1    0    0
```

Sample Problem 2-3: Matrix manipulation

Given are a 5×6 matrix A, a 3×6 matrix B, and a 9 element long vector v.

$$A = \begin{bmatrix} 2 & 5 & 8 & 11 & 14 & 17 \\ 3 & 6 & 9 & 12 & 15 & 18 \\ 4 & 7 & 10 & 13 & 16 & 19 \\ 5 & 8 & 11 & 14 & 17 & 20 \\ 6 & 9 & 12 & 15 & 18 & 21 \end{bmatrix} \qquad B = \begin{bmatrix} 5 & 10 & 15 & 20 & 25 & 30 \\ 30 & 35 & 40 & 45 & 50 & 55 \\ 55 & 60 & 65 & 70 & 75 & 80 \end{bmatrix}$$

$$v = \begin{bmatrix} 99 & 98 & 97 & 96 & 95 & 94 & 93 & 92 & 91 \end{bmatrix}$$

Create the three arrays in the Command Window, and then, by writing one command, replace the last four columns of the 1st and 3rd rows of A with the first four columns of the first two rows of B, the last four columns of the 4th row of A with the elements 5 through 8 of v, and the last four columns of the 5th row of A with columns 3 through 5 of the third row of B.

Solution

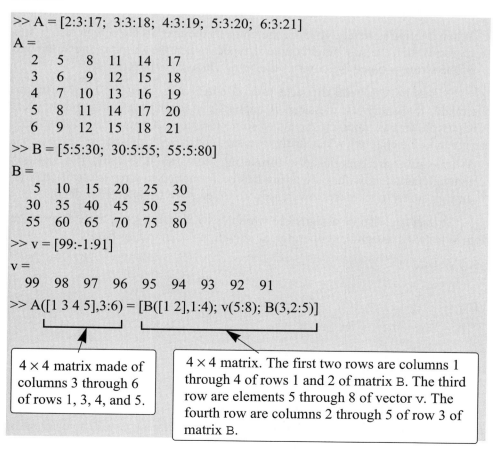

```
>> A = [2:3:17;  3:3:18;  4:3:19;  5:3:20;  6:3:21]
A =
    2    5    8    11    14    17
    3    6    9    12    15    18
    4    7   10    13    16    19
    5    8   11    14    17    20
    6    9   12    15    18    21
>> B = [5:5:30;  30:5:55;  55:5:80]
B =
    5   10   15   20   25   30
   30   35   40   45   50   55
   55   60   65   70   75   80
>> v = [99:-1:91]
v =
   99   98   97   96   95   94   93   92   91
>> A([1 3 4 5],3:6) = [B([1 2],1:4); v(5:8); B(3,2:5)]
```

4×4 matrix made of columns 3 through 6 of rows 1, 3, 4, and 5.	4×4 matrix. The first two rows are columns 1 through 4 of rows 1 and 2 of matrix B. The third row are elements 5 through 8 of vector v. The fourth row are columns 2 through 5 of row 3 of matrix B.

```
A =
    2    5    5   10   15   20
    3    6    9   12   15   18
    4    7   30   35   40   45
    5    8   95   94   93   92
    6    9   60   65   70   75
```

2.10 STRINGS AND STRINGS AS VARIABLES

- A string is an array of characters. It is created by typing the characters within single quotes.

- Strings can include letters, digits, other symbols, and spaces.

- Examples of strings: 'ad ef ', '3%fr2', '{edcba:21!', 'MATLAB'.

- A string that contains a single quote is created by typing two single quotes within the string.

- When a string is being typed in, the color of the text on the screen changes to maroon when the first single quote is typed. When the single quote at the end of the string is typed the color of the string changes to purple.

Strings have several different uses in MATLAB. They are used in output commands to display text messages (Chapter 4), in formatting commands of plots (Chapter 5), and as input arguments of some functions (Chapter 6). More details are given in the chapters when strings are used for these purposes.

- When strings are being used in formatting plots (labels to axes, title, and text notes), characters within the string can be formatted to have a specified font, size, position (uppercase, lowercase), color, etc. See Chapter 5 for details.

Strings can also be assigned to variables by simply typing the string on the right side of the assignment operator, as shown in the examples below:

```
>> a = 'FRty 8'

a =
FRty 8
>> B = 'My name is John Smith'

B =
My name is John Smith
>>
```

When a variable is defined as a string, the characters of the string are stored in an array just as numbers are. Each character, including a space, is an element in the array. This means that a one-line string is a row vector in which the number of elements is equal to the number of characters. The elements of the vectors are addressed by their position. For example, in the vector B that was defined above the fourth element is the letter n, the twelfth element is J and so on.

```
>> B(4)
ans =
n
>> B(12)
ans =
J
>>
```

As with a vector that contains numbers, it is also possible to change specific elements by addressing them directly. For example, in the vector B above the name John can be changed to Bill by:

```
>> B(12:15) = 'Bill'
B =
My name is Bill Smith
>>
```

> Using a colon to assign new characters to elements 12 through 15 in the vector B.

Strings can also be placed in a matrix. As with numbers, this is done by typing a semicolon ; (or pressing the **Enter** key) at the end of each row. Each row must be typed as a string, which means that it must be enclosed in single quotes. In addition, as with a numerical matrix, the number of elements in all the rows must be the same. This requirement can cause problems when the intention is to create rows with specific wording. Rows can be made to have the same number of elements by adding spaces.

MATLAB has a built-in function named char that creates an array with rows that have the same number of characters from an input of rows that are not of the same length. MATLAB makes the length of all the rows equal to the longest row by adding spaces at the end of the short lines. In the char function, the rows are entered as strings separated by a comma according to the following format:

```
variable_ name = char('string 1','string 2','string 3')
```

For example:

```
>> Info = char('Student Name:','John Smith','Grade:','A+')
Info =
Student Name:
John Smith
Grade:
A+
>>
```

> A variable named Info is assigned four rows of strings, each with different length.

> The function char creates an array with four rows with the same length as the longest row by adding empty spaces to the shorter lines.

A variable can be defined as a number or a string that is made up of the same digits. For example, as shown below, x is defined to be the number 536, and y is defined to be a string made up of the digits 536.

```
>> x = 536
x =
   536
>> y = '536'
y =
536
>>
```

The two variables are not the same even though they appear identical on the screen. The variable x can be used in mathematical expressions, while the variable y can not.

2.11 PROBLEMS

1. Create a row vector that has the elements: 32, 4, 81, $e^{2.5}$, 63, $\cos(\pi/3)$, and 14.12.

2. Create a column vector that has the elements: 55, 14, $\ln(51)$, 987, 0, and $5\sin(2.5\pi)$.

3. Create a row vector in which the first element is 1, the last element is 33, with an increment of 2 between the elements (1, 3, 5, ..., 33).

4. Create a column vector in which the first element is 15, the elements decrease with increments of –5, and the last element is –25. (A column vector can be created by the transpose of a row vector).

5. Create a row vector with 15 equally spaced elements in which the first element is 7 and the last element is 40.

6. Create a column vector with 12 equally spaced elements in which the first element is –1 and the last element is –15.

7. Create a vector, name it `Afirst`, that has 16 elements in which the first is 4, the increment is 3 and the last element is 49. Then, using the colon symbol, create a new vector, call it `Asecond`, that has eight elements. The first four elements are the first four elements of the vector `Afirst`, and the last four are the last four elements of the vector `Afirst`.

8. Create the matrix shown below by using the vector notation for creating vectors with constant spacing and/or the linspace command when entering the rows.

$$B = \begin{bmatrix} 1 & 4 & 7 & 10 & 13 & 16 & 19 & 22 & 25 \\ 72 & 66 & 60 & 54 & 48 & 42 & 36 & 30 & 24 \\ 0 & 0.125 & 0.250 & 0.375 & 0.500 & 0.625 & 0.750 & 0.875 & 1.000 \end{bmatrix}$$

9. Create the following matrix A: $\quad A = \begin{bmatrix} 6 & 43 & 2 & 11 & 87 \\ 12 & 6 & 34 & 0 & 5 \\ 34 & 18 & 7 & 41 & 9 \end{bmatrix}$

   Use the matrix A to:

   a) Create a five-element row vector named va that contains the elements of the second row of A.

   b) Create a three-element row vector named vb that contains the elements of the fourth column of A.

   c) Create a ten-element row vector named vc that contains the elements of the first and second rows of A.

   d) Create a six-element row vector named vd that contains the elements of the second and fifth columns of A.

10. Create the following matrix C: $\quad C = \begin{bmatrix} 2 & 4 & 6 & 8 & 10 \\ 3 & 6 & 9 & 12 & 15 \\ 7 & 14 & 21 & 28 & 35 \end{bmatrix}$

   Use the matrix C to:

   a) Create a three-element column vector named ua that contains the elements of the third column of C.

   b) Create a five-element column vector named ub that contains the elements of the second row of C.

   c) Create a nine-element column vector named uc that contains the elements of the first, third and fifth columns of C.

   d) Create a ten-element column vector named ud that contains the elements of the first and second rows of C.

11. Create the following matrix A: $\quad A = \begin{bmatrix} 1 & 2 & 3 & 4 & 5 & 6 & 7 \\ 2 & 4 & 6 & 8 & 10 & 12 & 14 \\ 21 & 18 & 15 & 12 & 9 & 6 & 3 \\ 5 & 10 & 15 & 20 & 25 & 30 & 35 \end{bmatrix}$

   a) Create a 3×4 matrix B from the 1st, 3rd, and 4th rows, and the 1st, 3rd, 5th, and 7th columns of the matrix A.

   b) Create a 15 element-long row vector u from the elements of the third row, and the 5th and 7th columns of the matrix A.

12. Using the zeros, ones, and eye commands create the following arrays:

a) $\begin{bmatrix} 0 & 0 & 0 & 0 & 0 \\ 0 & 0 & 0 & 0 & 0 \end{bmatrix}$
b) $\begin{bmatrix} 1 & 0 & 0 & 0 \\ 0 & 1 & 0 & 0 \\ 0 & 0 & 1 & 0 \\ 0 & 0 & 0 & 1 \end{bmatrix}$
c) $\begin{bmatrix} 1 & 1 \\ 1 & 1 \\ 1 & 1 \end{bmatrix}$

13. Using the eye command create the array A shown on the left below. Then, using the colon to address the elements in the array, change the array to be like the one shown on the right.

$$A = \begin{bmatrix} 1 & 0 & 0 & 0 & 0 & 0 & 0 \\ 0 & 1 & 0 & 0 & 0 & 0 & 0 \\ 0 & 0 & 1 & 0 & 0 & 0 & 0 \\ 0 & 0 & 0 & 1 & 0 & 0 & 0 \\ 0 & 0 & 0 & 0 & 1 & 0 & 0 \\ 0 & 0 & 0 & 0 & 0 & 1 & 0 \\ 0 & 0 & 0 & 0 & 0 & 0 & 1 \end{bmatrix} \qquad A = \begin{bmatrix} 2 & 2 & 2 & 0 & 5 & 5 & 5 \\ 2 & 2 & 2 & 0 & 5 & 5 & 5 \\ 3 & 3 & 3 & 0 & 5 & 5 & 5 \\ 0 & 0 & 0 & 1 & 0 & 0 & 0 \\ 4 & 4 & 7 & 0 & 9 & 9 & 9 \\ 4 & 4 & 7 & 0 & 9 & 9 & 9 \\ 4 & 4 & 7 & 0 & 9 & 9 & 9 \end{bmatrix}$$

14. Using the zeros and ones commands create a 3×5 matrix in which the first, second, and fifth columns are 0's, and the third and fourth columns are 1's.

15. Create a 5×7 matrix in which the first row are the numbers 1 2 3 4 5 6 7, the second row are the numbers 8 9 10 11 12 13 14, the third row are the numbers 15 through 21, and so on. From the this matrix create a new 3×4 matrix that is made from rows 2 through 4, and columns 3 through 6 of the first matrix.

16. Create a 3×3 matrix A in which all the elements are 1, and create a 2×2 matrix B in which all the elements are 5. Then, add elements to the matrix A by appending the matrix B such that A will be:

$$A = \begin{bmatrix} 1 & 1 & 1 & 0 & 0 \\ 1 & 1 & 1 & 0 & 0 \\ 1 & 1 & 1 & 0 & 0 \\ 0 & 0 & 0 & 5 & 5 \\ 0 & 0 & 0 & 5 & 5 \end{bmatrix}$$

Chapter 3
Mathematical Operations with Arrays

Once variables are created in MATLAB they can be used in a wide variety of mathematical operations. In Chapter 1 the variables that were used in mathematical operations were all defined as scalars. This means that they were all 1×1 arrays (arrays with one row and one column that have only one element) and the mathematical operations were done with single numbers. Arrays, however, can be one-dimensional (arrays with one row, or with one column), two-dimensional (arrays with rows and columns), and even of higher dimensions. In these cases the mathematical operations are more complex. MATLAB, as its name indicates, is designed to carry out advanced array operations that have many applications in science and engineering. This chapter presents the basic, most common mathematical operations that MATLAB performs using arrays.

Addition and subtraction are relatively simple operations and are covered first in Section 3.1. The other basic operations, multiplication, division and exponentiation, can be done in MATLAB in two different ways. One way, which uses the standard symbols (*, /, and ^), follows the rules of linear algebra and is presented in Sections 3.2 and 3.3. The second way, which is called element-by-element operations, is covered in Section 3.4. These operations use the symbols .*, ./, and .^ (a period is typed in front of the standard operation symbol). In addition, in both types of calculations, MATLAB has left division operators (.\ or \ ), which are also explained in Sections 3.3 and 3.4.

A note to first time users of MATLAB:

Although matrix operations are presented first and element-by-element operations next, the order can be reversed since the two are independent of each other. It is expected that almost every user has some knowledge of matrix operations and linear algebra, and thus will be able to follow the material covered in Sections 3.2 and 3.3 without any difficulty. Some readers, however, might prefer to read Section 3.4 first. MATLAB can be used with element-by-element operations in numerous applications that do not require linear algebra multiplication (or division) operations.

3.1 ADDITION AND SUBTRACTION

The operations + (addition) and − (subtraction) can be used with arrays of identical size (the same number of rows and columns). The sum, or the difference of two arrays is obtained by adding, or subtracting, their corresponding elements.

In general, if A and B are two arrays (for example 2×3 matrices),

$$A = \begin{bmatrix} A_{11} & A_{12} & A_{13} \\ A_{21} & A_{22} & A_{23} \end{bmatrix} \quad \text{and} \quad B = \begin{bmatrix} B_{11} & B_{12} & B_{13} \\ B_{21} & B_{22} & B_{23} \end{bmatrix}$$

then, the matrix that is obtained by adding A and B is:

$$\begin{bmatrix} (A_{11} + B_{11}) & (A_{12} + B_{12}) & (A_{13} + B_{13}) \\ (A_{21} + B_{21}) & (A_{22} + B_{22}) & (A_{23} + B_{23}) \end{bmatrix}$$

When a scalar (number) is added to, or subtracted from, an array, the number is added to, or subtracted from, all the elements of the array. Examples are:

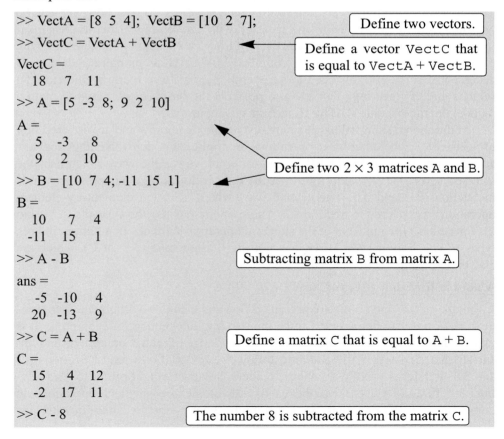

```
>> VectA = [8  5  4];  VectB = [10  2  7];
```
Define two vectors.
```
>> VectC = VectA + VectB
```
Define a vector VectC that is equal to VectA + VectB.
```
VectC =
    18    7   11
>> A = [5 -3 8; 9 2 10]
A =
     5   -3    8
     9    2   10
```
Define two 2 × 3 matrices A and B.
```
>> B = [10 7 4; -11 15 1]
B =
    10    7    4
   -11   15    1
>> A - B
```
Subtracting matrix B from matrix A.
```
ans =
    -5  -10    4
    20  -13    9
>> C = A + B
```
Define a matrix C that is equal to A + B.
```
C =
    15    4   12
    -2   17   11
>> C - 8
```
The number 8 is subtracted from the matrix C.

```
ans =
    7   -4   4
  -10    9   3
>>
```

3.2 ARRAY MULTIPLICATION

The multiplication operation * is executed by MATLAB according to the rules of linear algebra. This means that if A and B are two matrices, the operation $A*B$ can be carried out only if the number of columns in matrix A is equal to the number of rows in matrix B. The result is a matrix that has the same number of rows as A and the same number of columns as B. For example, if A is a 4×3 matrix and B is a 3×2 matrix:,

$$A = \begin{bmatrix} A_{11} & A_{12} & A_{13} \\ A_{21} & A_{22} & A_{23} \\ A_{31} & A_{32} & A_{33} \\ A_{41} & A_{42} & A_{43} \end{bmatrix} \text{ and } B = \begin{bmatrix} B_{11} & B_{12} \\ B_{21} & B_{22} \\ B_{31} & B_{32} \end{bmatrix}$$

then, the matrix that is obtained by the operation $A*B$ has the dimension of 4×2 with the elements:

$$\begin{bmatrix} (A_{11}B_{11} + A_{12}B_{21} + A_{13}B_{31}) & (A_{11}B_{12} + A_{12}B_{22} + A_{13}B_{32}) \\ (A_{21}B_{11} + A_{22}B_{21} + A_{23}B_{31}) & (A_{21}B_{12} + A_{22}B_{22} + A_{23}B_{32}) \\ (A_{31}B_{11} + A_{32}B_{21} + A_{33}B_{31}) & (A_{31}B_{12} + A_{32}B_{22} + A_{33}B_{32}) \\ (A_{41}B_{11} + A_{42}B_{21} + A_{43}B_{31}) & (A_{41}B_{12} + A_{42}B_{22} + A_{43}B_{32}) \end{bmatrix}$$

A numerical example is:

$$\begin{bmatrix} 1 & 4 & 3 \\ 2 & 6 & 1 \\ 5 & 2 & 8 \end{bmatrix} \begin{bmatrix} 5 & 4 \\ 1 & 3 \\ 2 & 6 \end{bmatrix} = \begin{bmatrix} (1 \cdot 5 + 4 \cdot 1 + 3 \cdot 2) & (1 \cdot 4 + 4 \cdot 3 + 3 \cdot 6) \\ (2 \cdot 5 + 6 \cdot 1 + 1 \cdot 2) & (2 \cdot 4 + 6 \cdot 3 + 1 \cdot 6) \\ (5 \cdot 5 + 2 \cdot 1 + 8 \cdot 2) & (5 \cdot 4 + 2 \cdot 3 + 8 \cdot 6) \end{bmatrix} = \begin{bmatrix} 15 & 34 \\ 18 & 32 \\ 43 & 74 \end{bmatrix}$$

The product of the multiplication of two square matrices (they must be of the same size) is also a square matrix of the same size. However, the multiplication of matrices is not commutative. This means that if A and B are both $n \times n$, then $A*B \neq B*A$. Also, the power operation can only be executed with a square matrix (since $A*A$ can be carried out only if the number of columns in the first matrix is equal to the number of rows in the second matrix).

Two vectors can multiply each other only if both have the same number of elements, and one is a row vector and the other is a column vector. The multiplication of a row vector times a column vector gives a 1×1 matrix, which is a scalar. This is the dot product of two vectors. (MATLAB also has a built-in function, named dot(a,b), that computes the dot product of two vectors.) When using the dot function, the vectors a and b can each be a row or a column vector, (see

Table 3-1). The multiplication of a column vector times a row vector, both with n elements gives an $n \times n$ matrix.

Tutorial 3-1: Multiplication of arrays.

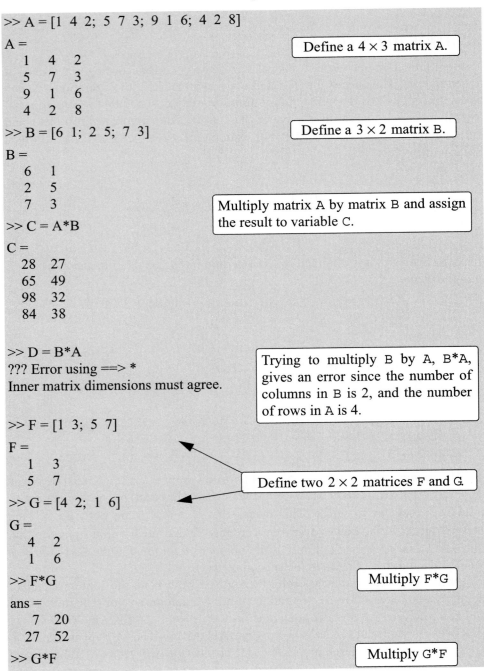

```
>> A = [1  4  2;  5  7  3;  9  1  6;  4  2  8]
A =
     1    4    2
     5    7    3
     9    1    6
     4    2    8
```
Define a 4 × 3 matrix A.

```
>> B = [6  1;  2  5;  7  3]
B =
     6    1
     2    5
     7    3
```
Define a 3 × 2 matrix B.

```
>> C = A*B
```
Multiply matrix A by matrix B and assign the result to variable C.

```
C =
    28   27
    65   49
    98   32
    84   38
```

```
>> D = B*A
??? Error using ==> *
Inner matrix dimensions must agree.
```
Trying to multiply B by A, B*A, gives an error since the number of columns in B is 2, and the number of rows in A is 4.

```
>> F = [1  3;  5  7]
F =
     1    3
     5    7
```
Define two 2 × 2 matrices F and G.

```
>> G = [4  2;  1  6]
G =
     4    2
     1    6
```

```
>> F*G
```
Multiply F*G

```
ans =
     7   20
    27   52
```

```
>> G*F
```
Multiply G*F

Tutorial 3-1: Multiplication of arrays. (Continued)

```
ans =
    14    26
    31    45
```
> Note that the answer for G*F is not the same as the answer for F*G.

```
>> AV = [2  5  1]
AV =
     2    5    1
```
> Define a three-element row vector AV.

```
>> BV = [3;  1;  4]
BV =
     3
     1
     4
```
> Define a three-element column vector BV.

```
>> AV * BV
ans =
    15
```
> Multiply AV by BV. The answer is a scalar. (Dot product of two vectors.)

```
>> BV * AV
ans =
     6    15    3
     2     5    1
     8    20    4
>>
```
> Multiply BV by AV. The answer is a 3×3 matrix.

When an array is multiplied by a number (actually a number is a 1×1 array), each element in the array is multiplied by the number. For example:

```
>> A = [2  5  7  0; 10 1 3 4; 6 2 11 5]
A =
     2    5    7    0
    10    1    3    4
     6    2   11    5
```
> Define a 3×4 matrix A.

```
>> b = 3
b =
     3
```
> Assign the number 3 to the variable b.

```
>> b*A
```
> Multiply the matrix A by b. This can be done by either typing b*A or A*b.

```
ans =
    6   15   21    0
   30    3    9   12
   18    6   33   15
>> C = A*5

C =
   10   25   35    0
   50    5   15   20
   30   10   55   25
>>
```

> Multiply the matrix A by 5 and assign the result to a new variable C.
> (Typing C = 5*A gives the same result.)

Linear algebra rules of array multiplication provide a convenient way for writing a system of linear equations. For example, the following system of three equations with three unknowns:

$$A_{11}x_1 + A_{12}x_2 + A_{13}x_3 = B_1$$
$$A_{21}x_1 + A_{22}x_2 + A_{23}x_3 = B_2$$
$$A_{31}x_1 + A_{32}x_2 + A_{33}x_3 = B_3$$

can be written in a matrix form by:

$$\begin{bmatrix} A_{11} & A_{12} & A_{13} \\ A_{21} & A_{22} & A_{23} \\ A_{31} & A_{32} & A_{33} \end{bmatrix} \begin{bmatrix} x_1 \\ x_2 \\ x_3 \end{bmatrix} = \begin{bmatrix} B_1 \\ B_2 \\ B_3 \end{bmatrix}$$

and in matrix notation by:

$$AX = B \quad \text{where } A = \begin{bmatrix} A_{11} & A_{12} & A_{13} \\ A_{21} & A_{22} & A_{23} \\ A_{31} & A_{32} & A_{33} \end{bmatrix}, X = \begin{bmatrix} x_1 \\ x_2 \\ x_3 \end{bmatrix}, \text{and } B = \begin{bmatrix} B_1 \\ B_2 \\ B_3 \end{bmatrix}.$$

3.3 ARRAY DIVISION

The division operation is also associated with the rules of linear algebra. This operation is more complex and only a brief explanation is given below. A full explanation can be found in books on linear algebra.

The division operation can be explained with the help of the identity matrix and the inverse operation.

Identity matrix:

The identity matrix is a square matrix in which the diagonal elements are 1's, and the rest of the elements are 0's. As was shown in Section 2.2.1, an identity matrix can be created in MATLAB with the eye command. When the identity matrix multiplies another matrix (or vector), that matrix (or vector) is unchanged (the multiplication has to be done according to the rules of linear algebra). This is equivalent to multiplying a scalar by 1. For example:

$$\begin{bmatrix} 7 & 3 & 8 \\ 4 & 11 & 5 \end{bmatrix}\begin{bmatrix} 1 & 0 & 0 \\ 0 & 1 & 0 \\ 0 & 0 & 1 \end{bmatrix} = \begin{bmatrix} 7 & 3 & 8 \\ 4 & 11 & 5 \end{bmatrix} \text{ or } \begin{bmatrix} 1 & 0 & 0 \\ 0 & 1 & 0 \\ 0 & 0 & 1 \end{bmatrix}\begin{bmatrix} 8 \\ 2 \\ 15 \end{bmatrix} = \begin{bmatrix} 8 \\ 2 \\ 15 \end{bmatrix} \text{ or } \begin{bmatrix} 6 & 2 & 9 \\ 1 & 8 & 3 \\ 7 & 4 & 5 \end{bmatrix}\begin{bmatrix} 1 & 0 & 0 \\ 0 & 1 & 0 \\ 0 & 0 & 1 \end{bmatrix} = \begin{bmatrix} 6 & 2 & 9 \\ 1 & 8 & 3 \\ 7 & 4 & 5 \end{bmatrix}$$

If a matrix A is square, it can be multiplied by the identity matrix, I, from the left or from the right:

$$AI = IA = A$$

Inverse of a matrix:

The matrix B is the inverse of the matrix A if when the two matrices are multiplied the product is the identity matrix. Both matrices must be square and the multiplication order can be BA or AB.

$$BA = AB = I$$

Obviously B is the inverse of A, and A is the inverse of B. For example:

$$\begin{bmatrix} 2 & 1 & 4 \\ 4 & 1 & 8 \\ 2 & -1 & 3 \end{bmatrix}\begin{bmatrix} 5.5 & -3.5 & 2 \\ 2 & -1 & 0 \\ -3 & 2 & 1 \end{bmatrix} = \begin{bmatrix} 5.5 & -3.5 & 2 \\ 2 & -1 & 0 \\ -3 & 2 & 1 \end{bmatrix}\begin{bmatrix} 2 & 1 & 4 \\ 4 & 1 & 8 \\ 2 & -1 & 3 \end{bmatrix} = \begin{bmatrix} 1 & 0 & 0 \\ 0 & 1 & 0 \\ 0 & 0 & 1 \end{bmatrix}$$

The inverse of a matrix A is typically written as A^{-1}. In MATLAB the inverse of a matrix can be obtained either by raising A to the power of -1, A^{-1}, or with the inv(A) function. For example, multiplying the matrices above with MATLAB is shown below:

```
>> A = [2  1  4;  4  1  8;  2 -1  3]
A =
     2     1     4
     4     1     8
     2    -1     3
```
Creating the matrix A.

```
>> B = inv(A)
B =
    5.5000   -3.5000    2.0000
    2.0000   -1.0000         0
   -3.0000    2.0000   -1.0000
```
Use the inv function to find the inverse of A and assign it to B.

```
>> A*B
ans =
     1     0     0
     0     1     0
     0     0     1
```
Multiplication of A and B gives the identity matrix.

```
>> A*A^-1

ans =
    1   0   0
    0   1   0
    0   0   1
```

Use power of -1 to find the inverse of A. Multiplying it by A gives the identity matrix.

- Not every matrix has an inverse. A matrix has an inverse only if it is square and its determinant is not equal to zero.

Determinants:

Determinant is a function associated with square matrices. A short review on determinants is given below. For a more detailed coverage the reader is referred to books on linear algebra.

The determinant is a function that associates with each square matrix A a number, called the determinant of the matrix. The determinant is typically denoted by $\det(A)$ or $|A|$. The determinant is calculated according to specific rules. For a second order 2×2 matrix the rule is:

$$A = \begin{vmatrix} a_{11} & a_{12} \\ a_{21} & a_{22} \end{vmatrix} = a_{11}a_{22} - a_{12}a_{21}, \text{ for example, } \begin{vmatrix} 6 & 5 \\ 3 & 9 \end{vmatrix} = 6 \cdot 9 - 5 \cdot 3 = 39$$

The determinant of a square matrix can be calculated with the det command (see Table 3-1).

Array division:

MATLAB has two types of array division, which are the right division and the left division.

Left division \ :

The left division is used to solve the matrix equation $AX = B$. In this equation X and B are column vectors. This equation can be solved by multiplying on the left both sides by the inverse of A:

$$A^{-1}AX = A^{-1}B$$

The left-hand side of this equation is X since:

$$A^{-1}AX = IX = X$$

So, the solution of $AX = B$ is:

$$X = A^{-1}B$$

In MATLAB the last equation can be written by using the left division character:

$$X = A \backslash B$$

It should be pointed out here that although the last two operations appear to give the same result, the method by which MATLAB calculates X is different. In the first, MATLAB calculates A^{-1} and then use it to multiply B. In the second, (left division) the solution X is obtained numerically with a method that is based on the Gauss elimination method. The left division method is recommended for solving a set of linear equations because the calculation of the inverse may be less accurate than the Gauss elimination method when large matrices are involved.

Right Division / :

The right division is used to solve the matrix equation $XC = D$. In this equation X and D are row vectors. This equation can be solved by multiplying on the right both sides by the inverse of C:

$$X \cdot CC^{-1} = D \cdot C^{-1}$$

which gives:

$$X = D \cdot C^{-1}$$

In MATLAB the last equation can be written by using the right division character:

$$X = D/C$$

The following example demonstrates the use of the left and right divisions, and the `inv` function to solve a set of linear equations.

Sample Problem 3-1: Solving three linear equations (array division)

Use matrix operations to solve the following system of linear equations.

$$4x - 2y + 6z = 8$$
$$2x + 8y + 2z = 4$$
$$6x + 10y + 3z = 0$$

Solution

Using the rules of linear algebra demonstrated earlier, the above system of equations can be written in the matrix form $AX = B$ or in the form $XC = D$:

$$\begin{bmatrix} 4 & -2 & 6 \\ 2 & 8 & 2 \\ 6 & 10 & 3 \end{bmatrix} \begin{bmatrix} x \\ y \\ z \end{bmatrix} = \begin{bmatrix} 8 \\ 4 \\ 0 \end{bmatrix} \quad \text{or} \quad \begin{bmatrix} x & y & z \end{bmatrix} \begin{bmatrix} 4 & 2 & 6 \\ -2 & 8 & 10 \\ 6 & 2 & 3 \end{bmatrix} = \begin{bmatrix} 8 & 4 & 0 \end{bmatrix}$$

The solution of both forms is shown below:

```
>> A = [4 -2 6; 2 8 2; 6 10 3];                    Solving the form AX = B.
>> B = [8; 4; 0];
>> X = A\B                          Solving by using left division X = A \ B.
X =
  -1.8049
   0.2927
   2.6341
```

```
>> Xb = inv(A)*B                          Solving by using the inverse of A  X = A⁻¹B.
Xb =
  -1.8049
   0.2927
   2.6341
>> C = [4 2 6; -2 8 10; 6 2 3];           Solving the form XC = D.
>> D = [8 4 0];
>> Xc = D/C                               Solving by using right division X = D/C.
Xc =
  -1.8049   0.2927   2.6341
>> Xd = D*inv(C)                          Solving by using the inverse of C,  X = D · C⁻¹.
Xd =
  -1.8049   0.2927   2.6341
```

Solving by using the inverse of A $X = A^{-1}B$.

Solving the form $XC = D$.

Solving by using right division $X = D/C$.

Solving by using the inverse of C, $X = D \cdot C^{-1}$.

3.4 ELEMENT-BY-ELEMENT OPERATIONS

In Sections 3.2 and 3.3 it was shown that when the regular symbols for multiplication and division are used with arrays (* and /), the mathematical operations follow the rules of linear algebra. There are, however, many situations that require element-by-element operations. These operations are carried out on each of the elements of the array (or arrays). Addition and subtraction are by definition already element-by-element operations since when two arrays are added (or subtracted) the operation is executed with the elements that are in the same position in the arrays. Element-by-element operations can only be done with arrays of the same size.

Element-by-element multiplication, division, and exponentiation of two vectors or matrices is entered in MATLAB by typing a period in front of the arithmetic operator.

Symbol	Description	Symbol	Description
.*	Multiplication	./	Right division
.^	Exponentiation	.\	Left Division

If two vectors a and b are: $a = \begin{bmatrix} a_1 & a_2 & a_3 & a_4 \end{bmatrix}$ and $b = \begin{bmatrix} b_1 & b_2 & b_3 & b_4 \end{bmatrix}$, then element-by-element multiplication, division, and exponentiation of the two vectors gives:

$$a .* b = \begin{bmatrix} a_1 b_1 & a_2 b_2 & a_3 b_3 & a_4 b_4 \end{bmatrix}$$

$$a ./ b = \begin{bmatrix} a_1/b_1 & a_2/b_2 & a_3/b_3 & a_4/b_4 \end{bmatrix}$$

$$a .^\wedge b = \left[(a_1)^{b_1} \ (a_2)^{b_2} \ (a_3)^{b_3} \ (a_4)^{b_4} \right]$$

If two matrices A and B are:

$$A = \begin{bmatrix} A_{11} & A_{12} & A_{13} \\ A_{21} & A_{22} & A_{23} \\ A_{31} & A_{32} & A_{33} \end{bmatrix} \quad \text{and} \quad B = \begin{bmatrix} B_{11} & B_{12} & B_{13} \\ B_{21} & B_{22} & B_{23} \\ B_{31} & B_{32} & B_{33} \end{bmatrix}$$

then element-by-element multiplication and division of the two matrices gives:

$$A .* B = \begin{bmatrix} A_{11}B_{11} & A_{12}B_{12} & A_{13}B_{13} \\ A_{21}B_{21} & A_{22}B_{22} & A_{23}B_{23} \\ A_{31}B_{31} & A_{32}B_{32} & A_{33}B_{33} \end{bmatrix} \qquad A ./ B = \begin{bmatrix} A_{11}/B_{11} & A_{12}/B_{12} & A_{13}/B_{13} \\ A_{21}/B_{21} & A_{22}/B_{22} & A_{23}/B_{23} \\ A_{31}/B_{31} & A_{32}/B_{32} & A_{33}/B_{33} \end{bmatrix}$$

Element-by-element exponentiation of matrix A gives:

$$A .^\wedge n = \begin{bmatrix} (A_{11})^n & (A_{12})^n & (A_{13})^n \\ (A_{21})^n & (A_{22})^n & (A_{23})^n \\ (A_{31})^n & (A_{32})^n & (A_{33})^n \end{bmatrix}$$

Element-by-element multiplication, division, and exponentiation are demonstrated in Tutorial 3-2.

Tutorial 3-2: Element-by-element operations.

```
>> A = [2  6  3;  5  8  4]
A =
    2    6    3
    5    8    4
```
Define a 2 × 3 array A.

```
>> B = [1  4  10;  3  2  7]
B =
    1    4   10
    3    2    7
```
Define a 2 × 3 array B.

```
>> A .* B

ans =
    2   24   30
   15   16   28
```
Element-by-element multiplication of array A by B.

```
>> C = A ./ B

C =
   2.0000   1.5000   0.3000
   1.6667   4.0000   0.5714
```
Element-by-element division of array A by B. The result is assigned to variable C.

Tutorial 3-2: Element-by-element operations. (Continued)

```
>> B .^ 3

ans =
        1      64    1000
       27       8     343
>> A * B

??? Error using ==> *
Inner matrix dimensions must agree.

>>
```

Element-by-element exponentiation of array B. The result is an array in which each term is the corresponding term in B raised to the power of 3.

Trying to multiply A*B gives an error since A and B cannot be multiplied according to linear algebra rules. (The number of columns in A is not equal to the number of rows in B.)

Element-by-element calculations are very useful for calculating the value of a function at many values of its argument. This is done by first defining a vector that contains values of the independent variable, and then using this vector in element-by-element computations to create a vector in which each element is the corresponding value of the function. One example is:

```
>> x = [1:8]
x =
     1   2   3   4   5   6   7   8
>> y = x.^2 - 4*x
y =
    -3  -4  -3   0   5  12  21  32
>>
```

Create a vector x with eight elements.

Vector x is used in element-by-element calculations of the elements of vector y.

In the example above $y = x^2 - 4x$. Element-by-element operation is needed when x is squared. Each element in the vector y is the value of y that is obtained when the value of the corresponding element of the vector x is substituted in the equation. Another example is:

```
>> z = [1:2:15]
z =
     1   3   5   7   9  11  13  15
>> y = (z.^3 + 5*z)./(4*z.^2 - 10)
```

Create a vector z with eight elements.

Vector z is used in element-by-element calculations of the elements of vector y.

```
y =
   -1.0000   1.6154   1.6667   2.0323   2.4650   2.9241   3.3964   3.8764
>>
```

In the example above $y = \dfrac{z^3 + 5z}{4z^2 - 10}$. Element-by-element operations are used in this example three times; to calculate z^3 and z^2, and to divide the numerator by the denominator.

3.5 USING ARRAYS IN MATLAB BUILT-IN MATH FUNCTIONS

The built-in functions in MATLAB are written such that when the argument (input) is an array, the operation that is defined by the function is executed on each element of the array. (One can think about the operation as element-by-element application of the function.) The result (output) from such an operation is an array in which each element is calculated by entering the corresponding element of the argument (input) array into the function. For example, if a vector with seven elements is substituted in the function cos (x), the result is a vector with seven elements in which each element is the cosine of the corresponding element in x. This is shown below:

```
>> x = [0:pi/6:pi]
x =
        0    0.5236    1.0472    1.5708    2.0944    2.6180    3.1416
>> y = cos(x)
y =
    1.0000    0.8660    0.5000    0.0000   -0.5000   -0.8660   -1.0000
>>
```

An example in which the argument variable is a matrix is:

```
>> d = [1 4 9; 16 25 36; 49 64 81]
d =
     1     4     9
    16    25    36
    49    64    81
>> h = sqrt(d)
h =
     1     2     3
     4     5     6
     7     8     9
>>
```

Creating a 3×3 array.

h is a 3×3 array in which each element is the square-root of the corresponding element in array d.

The feature of MATLAB, in which arrays can be used as arguments in functions, is called vectorization.

3.6 BUILT-IN FUNCTIONS FOR ANALYZING ARRAYS

MATLAB has many built-in functions for analyzing arrays. Table 3-1 lists some of these functions.

Table 3-1: Built-in array functions

Function	Description	Example
mean(A)	If A is a vector, returns the mean value of the elements of the vector.	>> A = [5 9 2 4]; >> mean(A) ans = 5
C=max(A)	If A is a vector, C is the largest element in A. If A is a matrix, C is a row vector containing the largest element of each column of A.	>> A = [5 9 2 4 11 6 7 11 0 1]; >> C = max(A) C = 11

[d,n]=max(A)	If A is a vector, d is the largest element in A, n is the position of the element (the first if several have the max value).	>> [d,n] = max(A)
		d =
		  11
		n =
		  5

min(A)	The same as max(A), but for the smallest element.	>> A = [5 9 2 4]; >> min(A) ans = 2
[d,n]=min(A)	The same as [d,n] = max(A), but for the smallest element.	
sum(A)	If A is a vector, returns the sum of the elements of the vector.	>> A = [5 9 2 4]; >> sum(A) ans = 20
sort(A)	If A is a vector, arranges the elements of the vector in ascending order.	>> A = [5 9 2 4]; >> sort(A) ans = 2 4 5 9
median(A)	If A is a vector, returns the median value of the elements of the vector.	>> A = [5 9 2 4]; >> median(A) ans = 4.5000

Table 3-1: Built-in array functions (Continued)

Function	Description	Example
std(A)	If A is a vector, returns the standard deviation of the elements of the vector.	>> A = [5 9 2 4]; >> std(A) ans = 2.9439
det(A)	Returns the determinant of a square matrix A.	>> A = [2 4; 3 5]; >> det(A) ans = -2
dot(a,b)	Calculates the scalar (dot) product of two vectors a and b. The vectors can each be row or column vectors.	>> a = [1 2 3]; >> b = [3 4 5]; >> dot(a,b) ans = 26
cross(a,b)	Calculates the cross product of two vectors a and b, (axb). The two vectors must have 3 elements.	>> a = [1 3 2]; >> b = [2 4 1]; >> cross(a,b) ans = -5 3 -2
inv(A)	Returns the inverse of a square matrix A.	>> A = [2 -2 1; 3 2 -1; 2 -3 2]; >> inv(A) ans = 0.2000 0.2000 0 -1.6000 0.4000 1.0000 -2.6000 0.4000 2.0000

3.7 GENERATION OF RANDOM NUMBERS

Simulations of many physical processes and engineering applications frequently requires using a number (or a set of numbers) that has a random value. MATLAB has two commands rand and randn that can be used to assign random numbers to variables.

The rand command:

The rand command generates uniformly distributed numbers with values between 0 and 1. The command can be used to assign these numbers to a scalar, a vector, or a matrix, as shown in Table 3-2.

Table 3-2: The `rand` command

Command	Description	Example
rand	Generates a single random number between 0 and 1.	>> rand ans = 0.2311
rand(1,n)	Generates an n element row vector of random numbers between 0 and 1.	>> a = rand(1,4) a = 0.6068 0.4860 0.8913 0.7621

rand(n)	Generates an n × n matrix with random numbers between 0 and 1.	>> b = rand(3)
		b =
		    0.4565 0.4447 0.9218
		    0.0185 0.6154 0.7382
		    0.8214 0.7919 0.1763

rand(m,n)	Generates an m × n matrix with random numbers between 0 and 1.	>> c = rand(2,4)
		c =
		    0.4057 0.9169 0.8936 0.3529
		    0.9355 0.4103 0.0579 0.8132

randperm(n)	Generates a row vector with n elements that are random permutation of integers 1 through n.	>> randperm(8) ans = 8 2 7 4 3 6 5 1

Sometimes there is a need to have random numbers that are distributed in an interval other than (0,1), or to have numbers that are only integers. This can be done by mathematical operations with the `rand` function. Random numbers that are distributed in a range (a,b) can be obtained by multiplying rand by $(b-a)$ and adding the product to a:

$$(b-a)*\texttt{rand} + a$$

For example, a vector of 10 elements with random values between -5 and 10 can be created by ($a = -5$, $b = 10$):

```
>> v = 15*rand(1,10) - 5
v =
  -1.8640    0.6973    6.7499    5.2127    1.9164    3.5174    6.9132   -4.1123
4.0430   -4.2460
```

Random numbers that are all integers can be generated by using one of the rounding functions. For example, a 2×15 matrix with random integers with values that range from 1 to 100 can be created by:

```
>> A = round(99*rand(2,15) + 1)
A =
   24   6  64  85  18  45  32  40  13  46  93  17  25  97  87
   25   9  20  18  99  35  37  60   5  87  27  87  65  67   2
```

The randn command:

The randn command generates normally distributed numbers with mean 0 and standard deviation of 1. The command can be used to generate a single number, a vector, or a matrix in the same way as the rand command. For example, a 3×4 matrix is created by:

```
>> d = randn(3,4)
d =
   -0.4326   0.2877   1.1892   0.1746
   -1.6656  -1.1465  -0.0376  -0.1867
    0.1253   1.1909   0.3273   0.7258
```

The mean and standard deviation of the numbers can be changed by mathematical operations to have any values. This is done by multiplying the number generated by the randn function by the desired standard deviation, and adding the desired mean. For example, a vector of 10 numbers (integers) with a mean of 50 and standard deviation of 5 is generated by:

```
>> v = round(5*randn(1,15) + 50)
v =
   53  51  45  39  50  45  53  53  58  53  47  52  45  50  50
```

In the example above integers are obtained by using the round function.

3.8 EXAMPLES OF MATLAB APPLICATIONS

Sample Problem 3-2:   Equivalent force system (addition of vectors)

Three forces are applied to a bracket as shown. Determine the total (equivalent) force applied to the bracket.

Solution

A force is a vector (physical quantity that has a magnitude and direction). In a Cartesian coordinate system a two-dimensional vector **F** can be written as:

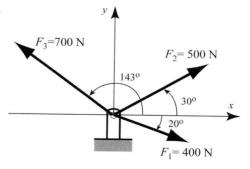

$$\mathbf{F} = F_x\mathbf{i} + F_y\mathbf{j} = F\cos\theta\mathbf{i} + F\sin\theta\mathbf{j} = F(\cos\theta\mathbf{i} + \sin\theta\mathbf{j})$$

where F is the magnitude of the force, and θ is its angle relative to the x axis, F_x and F_y are the components of $\mathbf{F}$ in the directions of the x and y axis, respectively, and $\mathbf{i}$ and $\mathbf{j}$ are unit vectors in these directions. If F_x and F_y are known, then F and θ can be determined by:

$$F = \sqrt{F_x^2 + F_y^2} \quad \text{and} \quad \tan\theta = \frac{F_y}{F_x}$$

The total (equivalent) force applied on the bracket is obtained by adding the forces that are acting on the bracket. The MATLAB solution below follows three steps:

- Write each force as a vector with two elements, where the first element is the x component of the vector and the second element is the y component.

- Determine the vector form of the equivalent force by adding the vectors.

- Determine the magnitude and direction of the equivalent force.

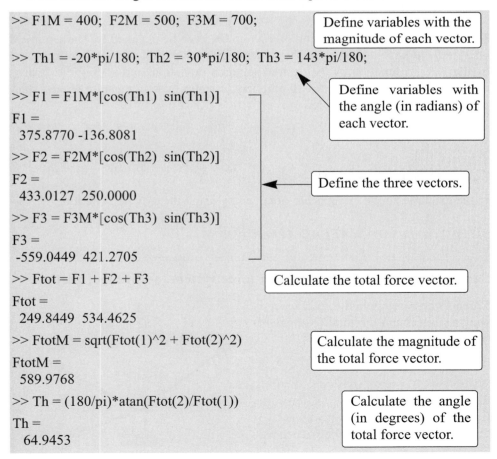

```
>> F1M = 400;  F2M = 500;  F3M = 700;
```
Define variables with the magnitude of each vector.

```
>> Th1 = -20*pi/180;  Th2 = 30*pi/180;  Th3 = 143*pi/180;
```

```
>> F1 = F1M*[cos(Th1)  sin(Th1)]
F1 =
  375.8770 -136.8081
```
Define variables with the angle (in radians) of each vector.

```
>> F2 = F2M*[cos(Th2)  sin(Th2)]
F2 =
  433.0127 250.0000
>> F3 = F3M*[cos(Th3)  sin(Th3)]
F3 =
 -559.0449 421.2705
```
Define the three vectors.

```
>> Ftot = F1 + F2 + F3
Ftot =
  249.8449 534.4625
```
Calculate the total force vector.

```
>> FtotM = sqrt(Ftot(1)^2 + Ftot(2)^2)
FtotM =
  589.9768
```
Calculate the magnitude of the total force vector.

```
>> Th = (180/pi)*atan(Ftot(2)/Ftot(1))
Th =
  64.9453
```
Calculate the angle (in degrees) of the total force vector.

The equivalent force has a magnitude of 504.62 N, and is directed 37.03° (ccw) relative to the x axis. In vector notation the force is: $\mathbf{F} = 402.83\mathbf{i} + 303.92\mathbf{j}$ N.

Sample Problem 3-3: Friction experiment (element-by-element calculations)

The coefficient of friction, μ, can be determined in an experiment by measuring the force F required to move a mass m. When F is measured and m is known, the coefficient of friction can be calculated by:

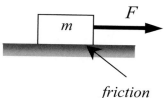

$$\mu = F/(mg) \quad (g = 9.81 \text{ m/s}^2).$$

friction

Results from measuring F in six tests are given in the table below. Determine the coefficient of friction in each test, and the average from all tests.

Test #	1	2	3	4	5	6
Mass m (kg)	2	4	5	10	20	50
Force F (N)	12.5	23.5	30	61	117	294

Solution

A solution using MATLAB commands in the Command Window is shown below.

```
>> m = [2  4 5  10 20  50];          Enter the values of m in a vector.
>> F = [12.5  23.5  30  61  117  294];   Enter the values of F in a vector.
>> mu = F./(m*9.81)
                              A value for mu is calculated for each test,
mu =                          using element-by-element calculations.


   0.6371   0.5989   0.6116   0.6218   0.5963   0.5994


>> mu_ave = mean(mu)
                              The average of the elements in the vector mu
mu_ave =                      is determined by using the function mean.

   0.6109
```

Sample Problem 3-4:   Electrical resistive network analysis (solving a system of linear equations)

The electrical circuit shown to the right consists of resistors and voltage sources. Determine the current in each resistor using the mesh current method which is based on Kirchhoff's voltage law.

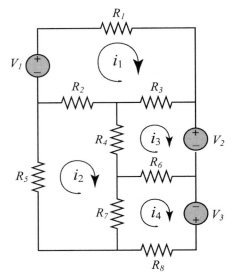

$V_1 = 20$ V, $V_2 = 12$ V, $V_3 = 40$ V
$R_1 = 18\,\Omega$, $R_2 = 10\,\Omega$, $R_3 = 16\,\Omega$
$R_4 = 6\,\Omega$, $R_5 = 15\,\Omega$, $R_6 = 8\,\Omega$
$R_7 = 12\,\Omega$, $R_8 = 14\,\Omega$

Solution

Kirchhoff's voltage law states that the sum of the voltage around a closed circuit is zero. In the mesh current method a current is first assigned for each mesh (i_1, i_2, i_3, i_4 in the figure). Then, Kirchhoff's voltage second law is applied for each mesh. This results in a system of linear equations for the currents (in our case four equations). The solution gives the values of the mesh currents. The current in a resistor that belongs to two meshes is the sum of the currents in the corresponding meshes. It is convenient to assume that all the currents are in the same direction (clockwise in our case). In the equation for each mesh, the voltage source is positive if the current flows to the – pole, and the voltage of a resistor is negative for current in the direction of the mesh current.

The equations for the four meshes in the current problem are:

$$V_1 - R_1 i_1 - R_3(i_1 - i_3) - R_2(i_1 - i_2) = 0$$
$$-R_5 i_2 - R_2(i_2 - i_1) - R_4(i_2 - i_3) - R_7(i_2 - i_4) = 0$$
$$-V_2 - R_6(i_3 - i_4) - R_4(i_3 - i_2) - R_3(i_3 - i_1) = 0$$
$$V_3 - R_8 i_4 - R_7(i_4 - i_2) - R_6(i_4 - i_3) = 0$$

The four equations can be rewritten in matrix form $[A][x] = [B]$:

$$\begin{bmatrix} -(R_1 + R_2 + R_3) & R_2 & R_3 & 0 \\ R_2 & -(R_2 + R_4 + R_5 + R_7) & R_4 & R_7 \\ R_3 & R_4 & -(R_3 + R_4 + R_6) & R_6 \\ 0 & R_7 & R_6 & -(R_6 + R_7 + R_8) \end{bmatrix} \begin{bmatrix} i_1 \\ i_2 \\ i_3 \\ i_4 \end{bmatrix} = \begin{bmatrix} -V_1 \\ 0 \\ V_2 \\ -V_3 \end{bmatrix}$$

A MATLAB solution of this system of equations is shown below:

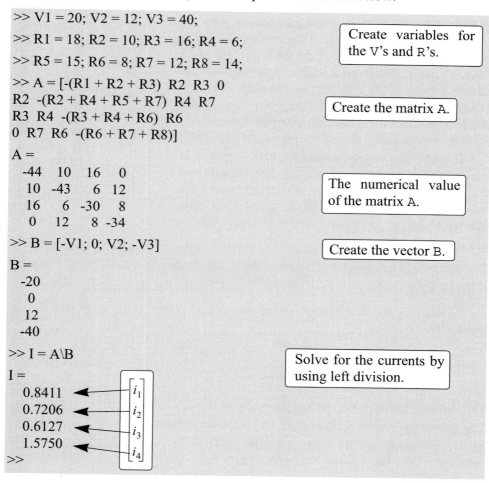

```
>> V1 = 20; V2 = 12; V3 = 40;
>> R1 = 18; R2 = 10; R3 = 16; R4 = 6;
>> R5 = 15; R6 = 8; R7 = 12; R8 = 14;
>> A = [-(R1 + R2 + R3)  R2  R3  0
R2  -(R2 + R4 + R5 + R7)  R4  R7
R3  R4  -(R3 + R4 + R6)  R6
0  R7  R6  -(R6 + R7 + R8)]
A =
   -44   10   16    0
    10  -43    6   12
    16    6  -30    8
     0   12    8  -34
>> B = [-V1; 0; V2; -V3]
B =
   -20
     0
    12
   -40
>> I = A\B
I =
    0.8411
    0.7206
    0.6127
    1.5750
>>
```

Create variables for the V's and R's.

Create the matrix A.

The numerical value of the matrix A.

Create the vector B.

Solve for the currents by using left division.

$\begin{bmatrix} i_1 \\ i_2 \\ i_3 \\ i_4 \end{bmatrix}$

The last column vector gives the current in each mesh. The currents in the resistors R_1, R_5, and R_8 are $i_1 = 0.8411$ A, $i_2 = 0.7206$ A, and $i_4 = 1.5750$ A, respectively. The other resistors belong to two meshes and their current is the sum of the currents in the meshes.

The current in resistor R_2 is $i_1 - i_2 = 0.1205$ A.

The current in resistor R_3 is $i_1 - i_3 = 0.2284$ A.

The current in resistor R_4 is $i_2 - i_3 = 0.1079$ A.

The current in resistor R_6 is $i_4 - i_3 = 0.9623$ A.

The current in resistor R_7 is $i_4 - i_2 = 0.8544$ A.

Sample Problem 3-5: Motion of two particles

A train and a car are approaching a road crossing. At $t = 0$ the train is 400 ft. south of the crossing traveling north at a constant speed of 54 mi/h. At the same time the car is 200 ft. west of the crossing traveling east at a speed of 28 mi/h and accelerating at 4 ft/s². Determine the positions of the train and the car, the distance between them, and the speed of the train relative to the car every second for the next 10 seconds.

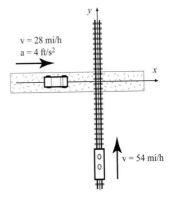

To show the results, create an 11×6 matrix in which each row has the time in the first column and the train position, car position, distance between the train and the car, car speed, and the speed of the train relative to the car, in the next five columns, respectively.

Solution

The position of an object that moves along a straight line at a constant acceleration is given by $s = s_o + v_o t + \frac{1}{2}at^2$ where s_o and v_o are the position and velocity at $t = 0$, and a is the acceleration. Applying this equation to the train and the car gives:

$$y = -400 + v_{otrain}t \quad \text{(train)}$$

$$x = -200 + v_{ocar}t + \frac{1}{2}a_{car}t^2 \quad \text{(car)}$$

The distance between the car and the train is: $d = \sqrt{x^2 + y^2}$.

The velocity of the train is constant and in vector notation is: $\mathbf{v}_{train} = v_{otrain}\mathbf{j}$. The car is accelerating and its velocity at time t is given by: $\mathbf{v}_{car} = (v_{ocar} + a_{car}t)\mathbf{i}$. The velocity of the train relative to the car, $\mathbf{v}_{t/c}$ is given by: $\mathbf{v}_{t/c} = \mathbf{v}_{train} - \mathbf{v}_{car} = -(v_{ocar} + a_{car}t)\mathbf{i} + v_{otrain}\mathbf{j}$. The magnitude (speed) of this velocity is the length of the vector.

    The problem is solved by first creating a vector t with 11 elements for the time from 0 to 10 s, and then calculating the positions of the train and the car, the distance between them, and the speed of the train relative to the car at each time element. The following are MATLAB commands that solve the problem.

```
>> v0train = 54*5280/3600; v0car = 28*5280/3600; acar = 4;
```
  Create variables for the initial velocities (in ft/s) and the acceleration.
```
>> t = 0:10;
```
  Create the vector t.
```
>> y = -400 + v0train*t;
```
```
>> x = -200 + v0car*t + 0.5*acar*t.^2;
```
  Calculate the train and car positions.
```
>> d = sqrt(x.^2 + y.^2);
```
  Calculate the distance between the train and car.

```
>> vcar = v0car + acar*t;                    [ Calculate the car's velocity. ]
>> speed_trainRcar = sqrt(vcar.^2 + v0train^2);   [ Calculate the speed of the
                                                    train relative to the car. ]
>> table = [t' y' x' d' speed_trainRcar']
                                              [ Create a table (see note below). ]
table =

        0 -400.0000 -200.0000 447.2136  41.0667   89.2139
   1.0000 -320.8000 -156.9333 357.1284  45.0667   91.1243
   2.0000 -241.6000 -109.8667 265.4077  49.0667   93.1675
   3.0000 -162.4000  -58.8000 172.7171  53.0667   95.3347
   4.0000  -83.2000   -3.7333  83.2837  57.0667   97.6178
   5.0000   -4.0000   55.3333  55.4777  61.0667  100.0089
   6.0000   75.2000  118.4000 140.2626  65.0667  102.5003
   7.0000  154.4000  185.4667 241.3239  69.0667  105.0849
   8.0000  233.6000  256.5333 346.9558  73.0667  107.7561
   9.0000  312.8000  331.6000 455.8535  77.0667  110.5075
  10.0000  392.0000  410.6667 567.7245  81.0667  113.3333
>>
```

Time (s)	Train position (ft)	Car position (ft)	Car-train distance (ft)	Car speed (ft/s)	Train speed relative to the car; (ft/s)

Note: In the commands above, `table` is the name of the variable that is a matrix containing the data to be displayed.

In this problem the results (numbers) are displayed by MATLAB without any text. Instructions on how to add text to output generated by MATLAB are presented in Chapter 4.

3.9 PROBLEMS

Note: *The problems below can be solved by writing commands in the Command Window. It may be more convenient, however, to use script files, which are explained in the next chapter. Please consider the possibility of first studying Chapter 4 (at least the first three sections), and then solving the following problems using script files. Additional problems for practicing mathematical operations with arrays are provided at the end of Chapter 4.*

1. Two vectors are given:

    $$\mathbf{u} = 4\mathbf{i} + 9\mathbf{j} - 5\mathbf{k} \quad \text{and} \quad \mathbf{v} = -3\mathbf{i} + 6\mathbf{j} - 7\mathbf{k}$$

    Use MATLAB to calculate the dot product $\mathbf{u} \cdot \mathbf{v}$ of the vectors in two ways:

    a) Define $\mathbf{u}$ as a row vector and $\mathbf{v}$ as a column vector, and then use matrix

    multiplication.
   b) Use the dot function.

2. For the function $y = (x^2 + 1)^3 x^3$, calculate the value of y for the following values of x: -2.5 -2 -1.5 -1 -0.5 0 0.5 1 1.5 2 2.5 3. Solve the problem by first creating a vector x, and then creating a vector y, using element-by-element calculations.

3. The depth of a well, d, in meters can be determined from the time it takes for a stone that is dropped into the well (zero initial velocity) to hit the bottom by: $d = \frac{1}{2}gt^2$, where t is the time in seconds and $g = 9.81$ m/s^2.

    Determine d for $t = 1,\ 2,\ 3,\ 4,\ 5,\ 6,\ 7,\ 8,\ 9$, and 10 s.
  (Create a vector t and determine d using element-by-element calculation.)

4. Define x and y as the vectors $x = 2, 4, 6, 8, 10$ and $y = 3, 6, 9, 12, 15$. Then use them in the following expression to calculate z using element-by-element calculations.

$$z = \frac{xy + \dfrac{y}{x}}{(x+y)^{(y-x)}} + 12^{x/y}$$

5. Define h and k as scalars, $h = 0.9$, and $k = 12.5$, and x, y and z as the vectors $x = 1, 2, 3, 4$, $y = 0.9, 0.8, 0.7, 0.6$, and $z = 2.5, 3, 3.5, 4$. Then use these variables to calculate T using element-by-element calculations for the vectors.

$$T = \frac{xyz}{(h+k)^{k/5}} + \frac{ke^{\left(\frac{z}{x}+y\right)}}{z^h}$$

6. Show that $\lim\limits_{n \to \infty}\left(1 + \dfrac{1}{n}\right)^n = e$

  Do this by first creating a vector n that has the elements: 1 10 100 500 1000 2000 4000 and 8000. Then, create a new vector y in which each element is determined from the elements of n by $\left(1 + \dfrac{1}{n}\right)^n$.

  Compare the elements of y with the value of e (type exp(1) to obtain the value of e).

7. Use MATLAB to show that the sum of the infinite series $\sum\limits_{n=1}^{\infty} \dfrac{1}{n^2}$ converges to $\pi^2/6$. Do it by computing the sum for:

  a) $n = 100$
  b) $n = 1{,}000$

c) $n = 10,000$

For each part create a vector n in which the first element is 1, the increment is 1 and the last term is 100, 1,000, or 10,000. Then, use element-by-element calculation to create a vector in which the elements are $1/n^2$. Finally, use the function sum to add the terms of the series. Compare the values obtained in parts *a*, *b*, and *c* with the value of $\frac{\pi^2}{6}$. (Don't forget to type semicolons at the end of commands that otherwise will display large vectors.)

8. Use MATLAB to show that the sum of the infinite series $\displaystyle\sum_{n=0}^{\infty} \frac{1}{(2n+1)(2n+2)}$ converges to *ln* 2. Do this by computing the sum for:
   a) $n = 50$
   b) $n = 500$
   c) $n = 5,000$

   For each part create a vector n in which the first element is 0, the increment is 1 and the last term is 50, 500, or 5,000. Then, use element-by-element calculation to create a vector in which the elements are $\dfrac{1}{(2n+1)(2n+2)}$. Finally, use the function sum to add the terms of the series. Compare the values obtained in parts *a*, *b*, and *c* to *ln* 2.

9. Create the following three matrices:

$$A = \begin{bmatrix} 5 & 2 & 4 \\ 1 & 7 & -3 \\ 6 & -10 & 0 \end{bmatrix} \qquad B = \begin{bmatrix} 11 & 5 & -3 \\ 0 & -12 & 4 \\ 2 & 6 & 1 \end{bmatrix} \qquad C = \begin{bmatrix} 7 & 14 & 1 \\ 10 & 3 & -2 \\ 8 & -5 & 9 \end{bmatrix}$$

   a) Calculate $A + B$ and $B + A$ to show that addition of matrices is commutative.

   b) Calculate $A + (B + C)$ and $(A + B) + C$ to show that addition of matrices is associative.

   c) Calculate $5(A + C)$ and $5A + 5C$ to show that, when matrices are multiplied by a scalar, the multiplication is distributive.

   d) Calculate $A*(B + C)$ and $A*B + A*C$ to show that matrix multiplication is distributive.

10. Use the matrices A, B, and C from the previous problem to answer the following:

    a) Does $A*B = B*A$?

    b)   Does $A*(B*C) = (A*B)*C$?

    c)   Does $(A*B)^t = B^t*A^t$? (^{t} means transpose)

    d)   Does $(A+B)^t = A^t + B^t$?

11. Solve the following system of four linear equations:

$$5x + 4y - 2z + 6w = 4$$
$$3x + 6y + 6z + 4.5w = 13.5$$
$$6x + 12y - 2z + 16w = 20$$
$$4x - 2y + 2z - 4w = 6$$

12. A projectile is fired with a muzzle speed of 750 m/s. Calculate the distance d that the projectile will hit the ground if the firing angle θ is changed from $5°$ to $85°$ in

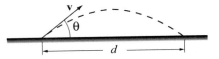

5° increments. Use element-by-element calculation. To show the results, create a 17×2 matrix in which the elements in the first column are the firing angles and the elements in the second column are the corresponding distances rounded to the nearest integers.

13. Two projectiles, A and B, are shot at the same instant from the same spot. Projectile A is shot at a speed of 680 m/s at an angle of $65°$ and projectile B is shot at a speed of 780 m/s at an angle of $42°$. Determine which projectile will hit the ground first. Then, take the flying time t_f of that projectile and divide it into ten increments by creating a vector t with 11 equally spaced elements (the first element is 0, the last is t_f). Calculate the distance between the two projectiles at the eleven times in vector t.

14. The electrical circuit shown consists of resistors and voltage sources. Determine the current in each resistor, using the mesh current method that is based on Kirchhoff's second voltage law (see Sample Problem 3-4).

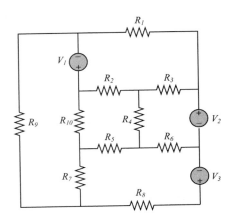

$V_1 = 38$ V, $V_2 = 20$ V, $V_3 = 24$ V
$R_1 = 15\,\Omega$, $R_2 = 18\,\Omega$, $R_3 = 10\,\Omega$
$R_4 = 9\,\Omega$, $R_5 = 5\,\Omega$, $R_6 = 14\,\Omega$
$R_7 = 8\,\Omega$, $R_8 = 13\,\Omega$, $R_9 = 5\,\Omega$
$R_{10} = 2\,\Omega$

Chapter 4
Script Files

So far in this book all the commands were executed in the Command Window. Although every MATLAB command can be executed in this way, using the Command Window to execute a series of commands—especially if they are related to each other (a program)—is not convenient and may be difficult or even impossible. The problem is that the commands in the Command Window cannot be saved and executed again. In addition, the Command Window is not interactive. This means that every time the **Enter** key is pressed only the last command is executed, and everything executed before is unchanged. If a change or a correction is needed in a command that was previously executed and the results of this command are used in commands that follow, all the commands have to be entered and executed again.

A different way of executing commands with MATLAB is first to create a file with a list of commands, save it, and then run the file. When the file runs, the commands it contains are executed in the order that they are listed. If needed, corrections or changes can be made to the commands in the file and the file can be saved and run again. The files that are used for this purpose are called script files.

4.1 NOTES ABOUT SCRIPT FILES

- A script file is a sequence of MATLAB commands, also called a program.

- When a script file runs, MATLAB executes the commands in the order they are written just as if they were typed in the Command Window.

- When a script file has a command that generates an output (e.g. assignment of a value to a variable without semicolon at the end), the output is displayed in the Command Window.

- Using a script file is convenient because it can be edited (corrected and/or changed) and executed many times.

- Script files can be typed and edited in any text editor and then pasted into the MATLAB editor.

- Script files are also called M-files because the extension .m is used when they are saved.

4.2 CREATING AND SAVING A SCRIPT FILE

In MATLAB script files are created and edited in the Editor/Debugger Window. This window is opened from the Command Window. In the **File** menu, select **New**, and then select **M-file**. An open Editor/Debugger Window is shown in Figure 4-1.

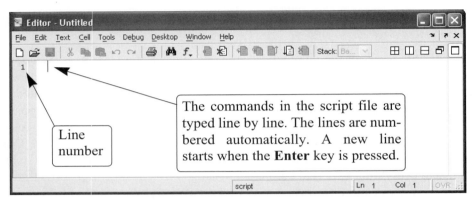

Figure 4-1: The Editor/Debugger Window.

Once the window is open, the commands of the script file are typed line by line. MATLAB automatically numbers a new line every time the **Enter** key is pressed. The commands can also be typed in any text editor or word processor program and then copied and pasted in the Editor/Debugger Window. An example of a short program typed in the Editor/Debugger Window is shown in Figure 4-2. The first few lines in a script file are typically comments (which are not executed since the first character in the line is %) that describe the program written in the script file.

Before a script file can be executed it has to be saved. This is done by choosing **Save As**... from the **File** menu, selecting a location (many students save to an A drive, or a zip drive), and entering a name for the file. The rules for naming a script file follow the rules of naming a variable (must begin with a letter, can include digits and underscore, and be up to 63 characters long). The names of user-defined variables, predefined variables, MATLAB commands or functions should not be used to name script files.

4.3 RUNNING A SCRIPT FILE

A script file can be executed either by typing its name in the Command Window and then pressing the **Enter** key, or directly from the Editor Window by clicking on the **Run** icon (see Figure 4-2). Before this can be done, however, the user has

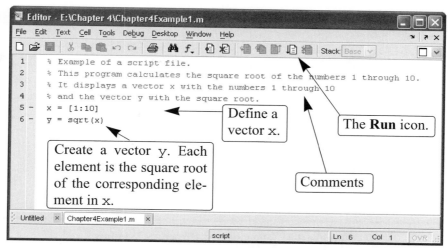

Figure 4-2: A program typed in the Editor/Debugger Window.

to make sure that MATLAB can find the file (MATLAB knows where the file is saved). In order to be able to run a file, the file must either be in the current directory, or in the search path.

4.3.1 Current directory

The current directory is shown in the "Current Directory" field in the desktop toolbar of the Command Window, as shown in Figure 4-3. The current directory

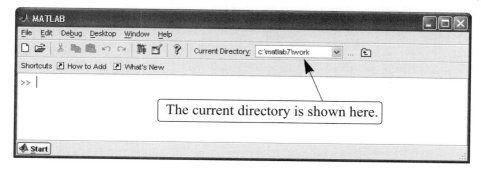

Figure 4-3: The Current Directory field in the Command Window.

can be changed either in the Current Directory Window, or by typing the cd command in the Command Window. Once two or more different current directories are used in a session, it is possible to switch from one to another in the **Current Directory** field in the Command Window. The Current Directory Window, shown in Figure 4-4, can be opened from the **Desktop** menu. The Current Directory can be changed by choosing the drive and folder where the file is saved.

An alternative simple way to change the current directory is to use the cd

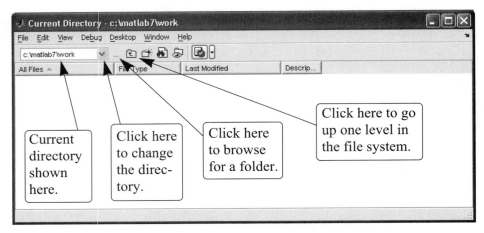

Figure 4-4: The Current Directory Window.

command in the Command Window. To change the current directory to a different drive, type cd, space, and then the name of the directory followed by a colon : and press the **Enter** key. For example, to change the current directory to drive A (usually the 3¹/₂ Floppy drive) type cd A:. If the script file is saved in a folder within a drive, the path to that folder has to be specified. This is done by typing the path as a string in the cd command. For example, cd('D:\Chapter 4') sets the path to the folder Chapter 4 in drive D. The following example shows how the current directory is changed to be drive A. Then, the script file from Figure 4-2, which was saved in drive A as: Chapter4Example1.m, is executed by typing the name of the file and pressing the **Enter** key.

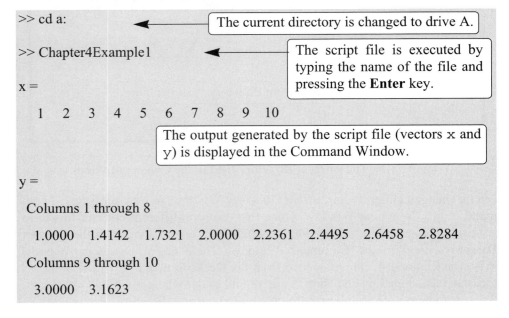

4.3.2 Search path

When MATLAB is asked to run a script file, or to execute a function, it searches for the file in directories that are listed in the search path. The directories that are included in the search path are displayed in the Set Path Window that can be opened by selecting **Set Path** in the **File** menu. Once the Set Path Window is open, new folders can be added to, or removed from, the search path.

4.4 GLOBAL VARIABLES

Global variables are variables that, once created in one part of MATLAB, are recognized in other parts of MATLAB. This is the case for variables in the Command Window and script files since both operate on variables in the workspace. When a variable is defined in the Command Window, it is also recognized and can be used in a script file. In the same way, if a variable is defined in a script file it is also recognized and can be used in the Command Window. In other words, once the variable is created, it exists, can be used, and can be reassigned a new value in both the Command Window and a script file.

There are different types of files in MATLAB, called function files, that normally don't share their variables with other parts of the program. This is explained later in Chapter 6 (Section 6.3).

4.5 INPUT TO A SCRIPT FILE

When a script file is executed the variables that are used in the calculations within the file must have assigned values. The assignment of a value to a variable can be done in three ways, depending on where and how the variable is defined.

1. The variable is defined and assigned value in the script file.

In this case the assignment of value to the variable is part of the script file. If the user wants to run the file with a different variable value, the file must be edited and the assignment of the variable changed. Then, after the file is saved, it can be executed again.

The following is an example of such a case. The script file (saved as Chapter4Example2) calculates the average points scored in three games.

```
% This script file calculates the average points scored in three games.
% The assignment of the values of the points is part of the script file.
game1 = 75;
game2 = 93;
game3 = 68;
ave_points = (game1 + game2 + game3)/3
```

The variables are assigned values within the script file.

The Command Window when this file is executed looks like:

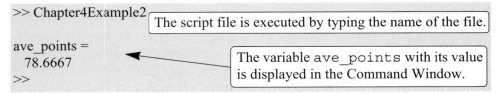

>> Chapter4Example2

| The script file is executed by typing the name of the file. |

ave_points =
   78.6667
>>

| The variable `ave_points` with its value is displayed in the Command Window. |

2. The variable is defined and assigned value in the Command Window.

In this case the assignment of a value to the variable is done in the Command Window. (Recall that the variable is recognized in the script file.) If the user wants to run the script file with a different value for the variable, the new value is assigned in the Command Window and the file is executed again.

For the previous example in which the script file has a program that calculates the average of points scored in three games, the script file (saved as Chapter4Example3) is:

% This script file calculates the average points scored in three games.

% The assignment of the values of the points to the variables

% game1, game2, and game3 is done in the Command Window.

ave_points = (game1+game2+game3)/3

The Command Window for running this file is:

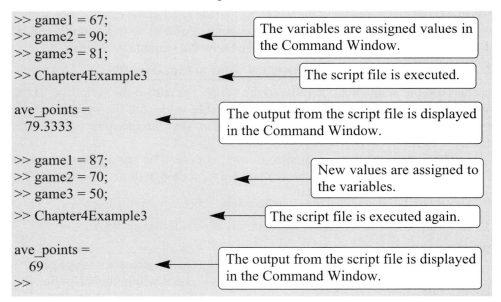

>> game1 = 67;
>> game2 = 90;
>> game3 = 81;

| The variables are assigned values in the Command Window. |

>> Chapter4Example3

| The script file is executed. |

ave_points =
   79.3333

| The output from the script file is displayed in the Command Window. |

>> game1 = 87;
>> game2 = 70;
>> game3 = 50;

| New values are assigned to the variables. |

>> Chapter4Example3

| The script file is executed again. |

ave_points =
   69
>>

| The output from the script file is displayed in the Command Window. |

3. The variable is defined in the script file, but a specific value is entered in the Command Window when the script file is executed.

In this case the variable is defined in the script file and when the file is executed, the user is prompted to assign a value to the variable in the Command Window. This is done by using the `input` command to create the variable.

The form of the `input` command is:

```
variable_name = input('string with a message that
                  is displayed in the Command Window')
```

When the `input` command is executed as the script file runs, the string is displayed in the Command Window. The string is a message prompting the user to enter a value that is assigned to the variable. The user types the value and presses the **Enter** key. This assigns the value to the variable. As with any variable, the variable and its assigned value will be displayed in the Command Window unless a semicolon is typed at the very end of the `input` command. A script file that uses the `input` command to enter the points scored in each game to the program that calculates the average of the scores is shown below.

```
% This script file calculates the average of points scored in three games.
% The point from each game are assigned to the variables by
% using the input command.
game1 = input('Enter the points scored in the first game   ');
game2 = input('Enter the points scored in the second game   ');
game3 = input('Enter the points scored in the third game   ');
ave_points = (game1 + game2 + game3)/3
```

The following shows the Command Window when this script file (saved as Chapter4Example4) is executed.

```
>> Chapter4Example4
Enter the points scored in the first game       67
Enter the points scored in the second game   91
Enter the points scored in the third game       70

ave_points =
    76
>>
```

> The computer displays the message. Then the value of the score is typed by the user and the **Enter** key is pressed.

In this example scalars are assigned to the variables. In general, however, vectors and arrays can also be assigned. This is done by typing the array in the same way that it is usually assigned to a variable (left bracket, then typing row by row, and a right bracket).

The input command can also be used to assign a string to a variable. This can be done in one of two ways. One way is to use the command in the same form as shown above, and when the prompt message appears the string is typed in between two single quotes in the same way that a string is assigned to a variable without the input command. The second way is to use an option in the input command that defines the characters that are entered as a string. The form of the command is:

```
variable_name = input('prompt message','s')
```

where the 's' inside the command defines the characters that will be entered as a string. In this case when the prompt message appears, the text is typed in without the single quotes, but it is assigned to the variable as a string. An example where the input command is used with this option is included in Sample Problem 7-4.

4.6 OUTPUT COMMANDS

As discussed before, MATLAB automatically generates a display when some commands are executed. For example, when a variable is assigned a value, or the name of a previously assigned variable is typed and the **Enter** key is pressed, MATLAB displays the variable and its value. This type of output is not displayed if a semicolon is typed at the end of the command. In addition to this automatic display, MATLAB has several commands that can be used to generate displays. The displays can be messages that provide information, numerical data, and plots. Two commands that are frequently used to generate output are the disp and fprintf. The disp command displays the output on the screen, while the fprintf command can be used to display the output on the screen, or to save the output to a file. The commands can be used in the Command Window, in a script file, and, as will be shown later, in a function file. When these commands are used in a script file, the display output that they generate is displayed in the Command Window.

4.6.1 The disp Command

The disp command is used to display the elements of a variable without displaying the name of the variable, and to display text. The format of the disp command is:

```
disp(name of a variable) or disp('text as string')
```

- Every time the disp command is executed, the display it generates appears in a new line. One example is:

```
>> abc = [5 9 1; 7 2 4];
```
A 2 × 3 array is assigned to variable abc.
```
>> disp(abc)
```
The disp command is used to display the abc array.
```
    5   9   1
    7   2   4
```
The array is displayed without its name.
```
>> disp('The problem has no solution.')

The problem has no solution.
>>
```
The disp command is used to display a message.

The next example shows the use of the disp command in the script file that calculates the average points scored in three games.

```
% This script file calculates the average points scored in three games.
% The points from each game are assigned to the variables by
% using the input command.
% The disp command is used to display the output.

game1 = input('Enter the points scored in the first game   ');
game2 = input('Enter the points scored in the second game   ');
game3 = input('Enter the points scored in the third game   ');
ave_points = (game1 + game2 + game3)/3;
disp(' ')
disp('The average of points scored in a game is:')
disp(' ')
disp(ave_points)
```
Display empty line.

Display text.

Display empty line.

Display the value of the variable ave_points.

When this file (saved as Chapter4Example5) is executed, the display in the Command Window is:

```
>> Chapter4Example5
Enter the points scored in the first game   89
Enter the points scored in the second game   60
Enter the points scored in the third game   82


The average of points scored in a game is:


   77
```
An empty line is displayed.

The text line is displayed.

An empty line is displayed.

The value of the variable ave_points is displayed.

- Only one variable can be displayed in a `disp` command. If elements of two variables need to be displayed together, a new variable (that contains the elements to be displayed) must first be defined and then displayed.

In many situations it is nice to display output (numbers) in a table. This can be done by first defining a variable that is an array with the numbers and then using the `disp` command to display the array. Headings to the columns can also be created with the `disp` command. Since in the `disp` command the user cannot control the format (the width of the columns and the distance between the columns) of the display of the array, the position of the headings has to be adjusted to the columns by adding spaces. As an example, the script file below shows how to display the population data from Chapter 2 in a table.

yr = [1984 1986 1988 1990 1992 1994 1996];   The population data is
pop = [127 130 136 145 158 178 211];   entered in two row vectors.
tableYP(:,1) = yr';   yr is entered as the first column in the array `tableYP`.
tableYP(:,2) = pop';   pop is entered as the second column in the array `tableYP`.
disp('   YEAR   POPULATION')   Display heading (first line).
disp('   (MILLIONS)')   Display heading (second line).
disp(' ')   Display an empty line.
disp(tableYP)   Display the array `tableYP`.

When this script file (saved as PopTable) is executed the display in the Command Window is:

```
>> PopTable
     YEAR    POPULATION          Headings are displayed.
              (MILLIONS)
                                 An empty line is displayed.

     1984       127
     1986       130
     1988       136              The tableYP array is displayed.
     1990       145
     1992       158
     1994       178
     1996       211
```

Another example of displaying a table is shown in Sample Problem 4-3. Tables can also be created and displayed with the `fprintf` command, which is explained in the next section.

4.6.2 The `fprintf` *Command*

The `fprintf` command can be used to display output (text and data) on the screen or to save it to a file. With this command (unlike with the `disp` command) the output can be formatted. For example, text and numerical values of variables can be intermixed and displayed in the same line. In addition, the format of the numbers can be controlled.

With many available options, the `fprintf` command can be long and complicated. To avoid confusion, the command is presented gradually. First, it is shown how to use the command to display text messages, then how to mix numerical data and text, followed by how to format the display of numbers, and lastly how to save the output to a file.

Using the `fprintf` command to display text:

To display text, the `fprintf` command has the form:

> ```
> fprintf('text typed in as a string')
> ```

For example:

> fprintf('The problem, as entered, has no solution. Please check the input data.')

If this line is part of a script file, when the line is executed, the following is displayed in the Command Window:

> The problem, as entered, has no solution. Please check the input data.

With the `fprintf` command it is possible to start a new line in the middle of the string. This is done by inserting `\n` before the character that will start the new line. For example, inserting `\n` after the first sentence in the previous example gives:

> fprintf('The problem, as entered, has no solution.\nPlease check the input data.')

When this line executes, the display in the Command Window is:

> The problem, as entered, has no solution.
> Please check the input data.

The `\n` is called an escape character. It is used to control the display. Other escape characters that can be inserted within the string are:

> `\b`   Backspace.
> `\t`   Horizontal tab.

When a program has more than one `fprintf` command, the display that they generate is continuous (the `fprintf` command does not automatically start a new line). This is true even if there are other commands between the `fprintf` commands. An example is the following script file:

```
fprintf('The problem, as entered, has no solution. Please check the input data.')
x = 6; d = 19 + 5*x;
fprintf('Try to run the program later.')
y = d + x;
fprintf('Use different input values.')
```

When this file is executed the display in the Command Window is:

The problem, as entered, has no solution. Please check the input data.Try to run the program later.Use different input values.

To start a new line with the `fprintf` command, a `\n` must be typed at the start of the string.

Using the `fprintf` command to display a mix of text and numerical data:

To display a mix of text and a number (value of a variable), the `fprintf` command has the form:

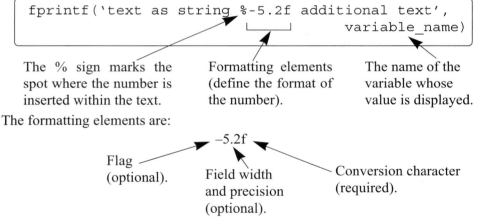

The % sign marks the spot where the number is inserted within the text.

Formatting elements (define the format of the number).

The name of the variable whose value is displayed.

The formatting elements are:

Flag (optional).

Field width and precision (optional).

Conversion character (required).

The flag, which is optional can be one of the following three:

Character used for flag	Description
– (minus sign)	Left justifies the number within the field.
+ (plus sign)	Prints a sign character (+ or –) in front of the number.
0 (zero)	Adds zeros if the number is shorter than the field.

The field width and precision (5.2 in the previous example) are optional. The first number (5 in the example) is the field width that specifies the minimum number of digits in the display. If the number to be displayed is shorter than the field width, spaces or zeros are added in front of the number. The precision is the second number (2 in the previous example) specifies the number of digits to be displayed to the right of the decimal point.

The last element in the formatting elements, which is required, is the conversion character which specifies the notation in which the number is displayed. Some of the common notations are:

e   Exponential notation using lower case e (e.g. 1.709098e+001).
E   Exponential notation using upper case E (e.g. 1.709098E+001).
f   Fixed point notation (e.g. 17.090980).
g   The shorter of e or f notations.
G   The shorter of E or f notations.
i   Integer.

Information about additional notation is available in the help menu of MATLAB. As an example, the `fprintf` command with a mix of text and a number is used in the script file that calculates the average points scored in three games.

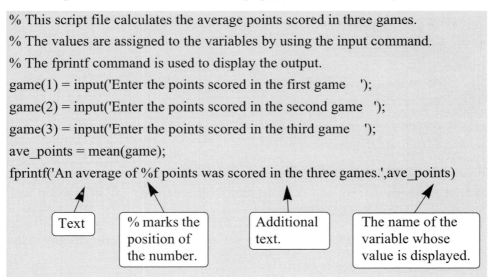

```
% This script file calculates the average points scored in three games.
% The values are assigned to the variables by using the input command.
% The fprintf command is used to display the output.
game(1) = input('Enter the points scored in the first game   ');
game(2) = input('Enter the points scored in the second game   ');
game(3) = input('Enter the points scored in the third game   ');
ave_points = mean(game);
fprintf('An average of %f points was scored in the three games.',ave_points)
```

| Text | % marks the position of the number. | Additional text. | The name of the variable whose value is displayed. |

Notice that, besides using the `fprintf` command, this file differs from the ones shown earlier in the chapter in that the scores are stored in the first three elements of a vector named `game`, and the average of the scores is calculated by using the `mean` function. The Command Window where the script file above (saved as Chapter4Example6) was run is shown below.

```
>> Chapter4Example6
Enter the points scored in the first game   75
Enter the points scored in the second game   60
Enter the points scored in the third game   81
```

An average of 72.000000 points was scored in the three games.

>>
The display generated by the `fprintf` command combines text and a number (value of a variable).

With the `fprintf` command it is possible to insert more than one number (values of a variable) within the text. This is done by typing %g (or % followed by any formatting elements) at the places in the text where the numbers are to be inserted. Then, after the string argument of the command (following the comma), the names of the variables are typed in the order that they are inserted in the text. In general the command looks like:

```
fprintf('..text...%g...%g...%f...',variable1,variable2,variable3)
```

An example is shown in the following script file:

% This program calculates the distance a projectile flies,

% given its initial velocity and the angle at which it is shot.

% the fprintf command is used to display a mix of text and numbers.

v = 1584; % Initial velocity (km/h)

theta = 30; % Angle (degrees)

vms = v*1000/3600;   Changing velocity units to m/s.

t = vms*sin(30*pi/180)/9.81;   Calculating the time to highest point.

d = vms*cos(30*pi/180)*2*t/1000;   Calculating max distance.

fprintf('A projectile shot at %3.2f degrees with a velocity of %4.2f km/h will travel a distance of %g km.',theta,v,d)

When this script file (saved as Chapter4Example7) is executed, the display in the Command Window is:

>> Chapter4Example7

A projectile shot at 30.00 degrees with a velocity of 1584.00 km/h will travel a distance of 17.091 km.

>>

Additional Remarks About the `fprintf` Command:

- To have a single quotation mark in the displayed text, type two single quotation marks in the string inside the command.

- The `fprintf` command is vectorized. This means that when a variable that is a vector or a matrix is included in the command, the command repeats itself until all the elements are displayed. If the variable is a matrix the data is used column by column.

For example, the script file below creates a 2×5 matrix T in which the first row are the numbers 1 through 5, and the second row are the corresponding square roots.

```
x = 1:5;
y = sqrt(x);
T = [x; y]
fprintf('If the number is: %i, its square root is: %f\n',T)
```

Create a vector x.

Create a vector y.

Create 2×5 matrix T, first row is x, second row is y.

The fprintf command displays two numbers from T in every line.

When this script file is executed the display in the Command Window is:

```
T =
   1.0000   2.0000   3.0000   4.0000   5.0000
   1.0000   1.4142   1.7321   2.0000   2.2361
If the number is: 1, its square root is: 1.000000
If the number is: 2, its square root is: 1.414214
If the number is: 3, its square root is: 1.732051
If the number is: 4, its square root is: 2.000000
If the number is: 5, its square root is: 2.236068
```

The 2×5 matrix T.

The fprintf command repeats 5 times, using the numbers from the matrix T column after column.

Using the fprintf command to save output to a file:

In addition to displaying output in the Command Window, the fprintf command can be used for writing the output to a file when it is necessary to save the output. The data that is saved can subsequently be displayed or used in MATLAB and in other applications.

Writing output to a file requires three steps:

  a) Opening a file using the fopen command.
  b) Writing the output to the open file using the fprintf command.
  c) Closing the file using the fclose command.

Step *a*:

Before data can be written to a file, the file must be opened. This is done with the fopen command, which creates a new file or opens an existing file. The fopen command has the form:

```
fid = fopen('file_name','permission')
```

fid is a variable called the file identifier. A scalar value is assigned to fid when fopen is executed. The file name is written (including its extension) within single quotes as a string. The permission is a code (also written as a string) that tells how the file is opened. Some of the more common permission codes are:

'r'	Open file for reading (default).
'w'	Open file for writing. If the file already exists, its content is deleted. If the file does not exists, a new file is created.
'a'	Same as 'w', except that if the file exists the written data is appended to the end of the file.

If a permission code is not included in the command, the file opens with the default code 'r'. Additional permission codes are described in the help menu.

Step *b*:

Once the file is open, the fprintf command can be used to write output to the file. The fprintf command is used in exactly the same way as it is used to display output in the Command Window, except that the variable fid is inserted inside the command. The fprintf command then has the form:

```
fprintf(fid,'text %-5.2f additional text',vari
                                       able_name)
```

fid is added to the fprintf command.

Step *c*:

When writing the data to the file is complete, the file is closed using the fclose command. The fclose command has the form:

```
fclose(fid)
```

Additional notes on using the fprintf command for saving output to a file:

- The created file is saved in the current directory.

- It is possible to use the fprintf command to write to several different files. This is done by first opening the files, assigning a different fid to each (e.g. fid1, fid2, fid3, etc.), and then using the fid of a specific file in the fprintf command to write to that file.

An example of using fprintf commands for saving output to two files is shown in the following script file. The program in the file generates two unit conversion tables. One table converts velocity units from miles per hour to kilometers per hour, and the other table converts force units from pounds to Newtons. Each conversion table is saved to a different text file (extension .txt).

% Script file in which fprintf is used to write output to files.

% Two conversion tables are created and saved to two different files.

```
% One converts mi/h to km/h, the other converts lb to N.
clear all
Vmph = 10:10:100;
Vkmh = Vmph.*1.609;
TBL1 = [Vmph; Vkmh];
Flb = 200:200:2000;
FN = Flb.*4.448;
TBL2 = [Flb; FN];
fid1 = fopen('VmphtoVkm.txt','w');
fid2 = fopen('FlbtoFN.txt','w');
fprintf(fid1,'Velocity Conversion Table\n \n');

fprintf(fid1,'    mi/h        km/h  \n');

fprintf(fid1,'  %8.2f    %8.2f\n',TBL1);

fprintf(fid2,'Force Conversion Table\n \n');
fprintf(fid2,'  Pounds        Newtons  \n');
fprintf(fid2,'  %8.2f     %8.2f\n',TBL2);
fclose(fid1);
fclose(fid2);
```

Creating a vector of velocities in mph.

Converting mph to km/h.

Creating a table (matrix) with two rows.

Creating a vector of forces in lb.

Converting lb to N.

Creating a table (matrix) with two rows.

Open a txt file named VmphtoVkm.

Open a txt file named FlbtoFN.

Writing a title and an empty line to the file fid1.

Writing two columns heading to the file fid1.

Writing the data from the variable TBL1 to the file fid1.

Writing the force conversion table (data in variable TBL2) to the file fid2.

Closing the files fid1 and fid2.

When the script file above is executed two new txt files, named Vmph-

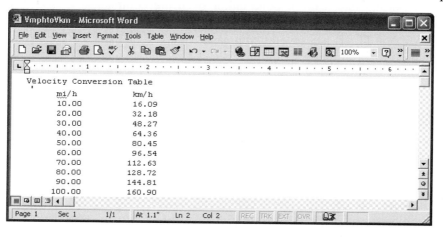

Figure 4-5: The Vmphtokm.txt file opened in Word.

toVkm and FlbtoFN, are created and saved in the current directory. These files can be opened with any application that can read txt files. Figures 4-5 and 4-6 show how the two files appear when they are opened with Microsoft Word.

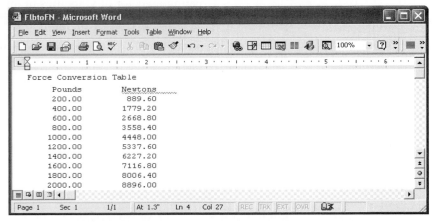

Figure 4-6: The FlbtoFN.txt file opened in Word.

4.7 IMPORTING AND EXPORTING DATA

MATLAB is often used for analyzing data that was recorded in experiments or generated by other computer programs. This can be done by first importing the data into MATLAB. Similarly, data that is produced by MATLAB sometimes needs to be transferred to other computer applications. There are various types of data (numerical, text, audio, graphics, and images). This section only describes how to import and export numerical data which is probably the most common type of data that need to be transferred by new users of MATLAB. For other types of data transfer, the user should look in the Help Window under File I/O.

Importing data can be done either by using commands, or by using the Import Wizard. Commands are useful when the format of the data being imported is known. MATLAB has several commands that can be used for importing various types of data. Importing commands can also be included in a script file such that the data is imported when the script is executed. The Import Wizard is useful when the format of the data (or the command that is applicable for importing the data) is not known. The Import Wizard determines the format of the data and automatically imports it.

4.7.1 Commands for Importing and Exporting Data

This section describes – in detail – how to transfer data into and out of Excel spreadsheets. Microsoft Excel is commonly used for storing data, and Excel is compatible with many data recording devices and computer applications. Many people are also capable of importing and exporting various data formats into and from Excel. MATLAB has also commands for transferring data directly to and from formats like csv and ASCII, and to the spreadsheet program Lotus 123. Details of these and many other commands can be found in the Help Window under File I/O

Importing and exporting data into and from Excel:

Importing data from Excel is done with the `xlsread` command. When the command is executed, the data from the spreadsheet is assigned as an array to a variable. The simplest form of the `xlsread` command is:

> `variable_name = xlsread('filename')`

- `'filename'` (typed as a string) is the name of the Excel file. The directory of the Excel file must either be the current directory, or listed in the search path.

- If the Excel file has more than one sheet, the data will be imported from the first sheet.

When an Excel file has several sheets, the `xlsread` command can be used to import data from a specified sheet. The form of the command is then:

> `variable_name = xlsread('filename','sheet_name')`

- The name of the sheet is typed as a string.

Another option is to import only a portion of the data that is in the spreadsheet. This is done by typing an additional argument in the command:

> `variable_name = xlsread('filename','sheet_name','range')`

- The `'range'` (typed as a string) is a rectangular region of the spreadsheet defined by the addresses (in Excel notation) of the cells at opposite corners of the region. For example, `'C2:E5'` is a 4 by 3 region of rows 2, 3, 4, and 5 and columns C, D, and E.

Exporting data from MATLAB to an Excel spreadsheet is done by using the `xlswrite` command. The simplest form of the command is:

> `xlswrite('filename',variable_name)`

- `'filename'` (typed as a string) is the name of the Excel file to which the data is exported. The file must be in the current directory. If the file does not exist, a new Excel file with the specified name will be created.

- `variable_name` is the name of the variable in MATLAB with the assigned data that is being exported.

- The arguments `'sheet_name'` and `'range'` can be added to the `xlswrite` command to export to a specified sheet and to a specified range of cells, respectively.

As an example, the data from the Excel spreadsheet shown in Figure 4-7 is imported into MATLAB by using the `xlsread` command.

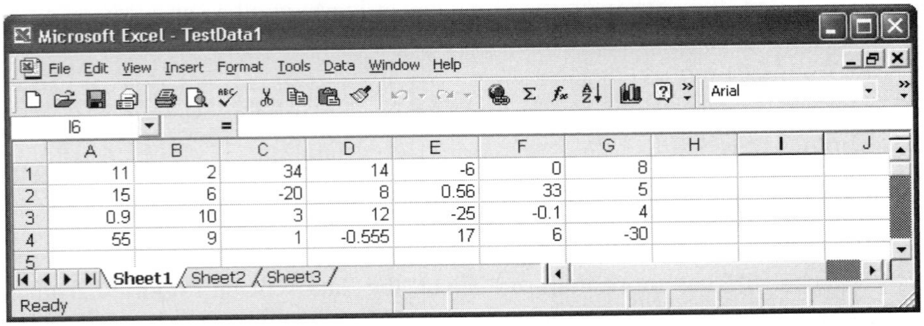

Figure 4-7: Excel spreadsheet with data.

The spreadsheet is saved in a file named TestData1 in a disk in drive A. After the Current Directory is changed to drive A, the data is imported into MATLAB by assigning it to the variable DATA:

```
>> DATA = xlsread('TestData1')
DATA =
   11.0000    2.0000   34.0000   14.0000   -6.0000        0    8.0000
   15.0000    6.0000  -20.0000    8.0000    0.5600  33.0000    5.0000
    0.9000   10.0000    3.0000   12.0000  -25.0000   -0.1000    4.0000
   55.0000    9.0000    1.0000   -0.5550   17.0000    6.0000  -30.0000
```

4.7.2 Using the Import Wizard

Using the Import Wizard is probably the easiest way to import data into MATLAB since the user does not have to know, or to specify, the format of the data. The Import Wizard is activated by selecting **Import Data** in the **File** menu of the Command Window. (It can also be started by typing the command `uiimport`.) The Import Wizard starts by displaying a file selection box that shows all the data files recognized by the Wizard. The user then selects the file which contains the data to be imported, and clicks **Open**. The Import Wizard opens the file and displays a portion of the data in a preview box of the wizard such that the user can verify that the data is the correct one. The Import Wizard tries to process the data, and if the wizard is successful, it displays the variables it has created with a portion of the data. The user clicks **next** and the wizard shows the Column Separator that was used. If the variable has the correct data, the user can proceed with the wizard (click **next**); otherwise the user can choose a different Column Separator.

In the next window the wizard shows the name and size of the variable to be created in MATLAB. (When the data is all numerical, the variable in MATLAB has the same name as the file from which the data was imported.) When the wizard ends (click **finish**), the data is imported to MATLAB.

As an example, the Import Wizard is used to import numerical ASCII data saved in a txt file. The data saved with the file name TestData2 is shown in Figure 4-8.

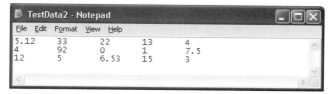

Figure 4-8: Numerical ASCII data.

The display of the Import Wizard during the import process of the TestData2 file is shown in Figures 4-9 through 4-11. The third figure shows that the name of the

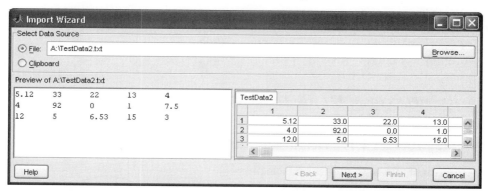

Figure 4-9: Import Wizard, first display.

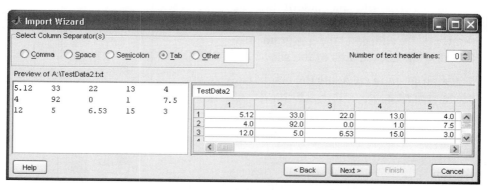

Figure 4-10: Import Wizard, second display.

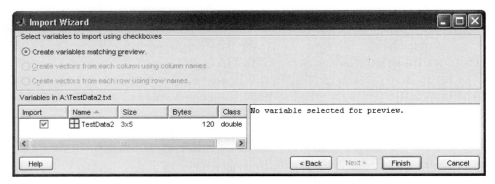

Figure 4-11: Import Wizard, third display.

variable in MATLAB is TestData2, and its size is 3 by 5. In the Command Window of MATLAB, the imported data can be displayed by typing the name of the variable.

```
>> TestData2

TestData2 =
     5.1200   33.0000   22.0000   13.0000    4.0000
     4.0000   92.0000         0    1.0000    7.5000
    12.0000    5.0000    6.5300   15.0000    3.0000
```

4.8 EXAMPLES OF MATLAB APPLICATIONS

Sample Problem 4-1:   Height and surface area of a silo

A cylindrical silo with radius r has a spherical cap roof with radius R. The height of the cylindrical portion is H. Write a program in a script file that determines the height H for given values of r, R, and the volume V. In addition, the program also calculates the surface area of the silo.

Use the program to calculate the height and surface area of a silo with $r = 30$ ft., $R = 45$ ft., and a volume of 120,000 ft^3. Assign values for r, R, and V in the Command Window.

Solution

The total volume of the silo is obtained by adding the volume of the cylindrical part and the volume of the spherical cap. The volume of the cylinder is given by:

$$V_{cyl} = \pi r^2 H$$

and the volume of the spherical cap is given by:

$$V_{cap} = \frac{1}{3}\pi h^2 (3R - h)$$

where $h = R - R\cos\theta = R(1 - \cos\theta)$,

and θ is calculated from $\sin\theta = \frac{r}{R}$.

Using the equations above, the height, H, of the cylindrical part can be expressed by:

$$H = \frac{V - V_{cap}}{\pi r^2}$$

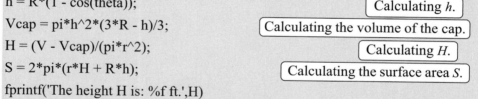

The surface area of the silo is obtained by adding the surface areas of the cylindrical part and the spherical cap.

$$S = S_{cyl} + S_{cap} = 2\pi rH + 2\pi Rh$$

A program in a script file that solves the problem is presented below:

```
theta = asin(r/R);                  Calculating θ.
h = R*(1 - cos(theta));             Calculating h.
Vcap = pi*h^2*(3*R - h)/3;          Calculating the volume of the cap.
H = (V - Vcap)/(pi*r^2);            Calculating H.
S = 2*pi*(r*H + R*h);              Calculating the surface area S.
fprintf('The height H is: %f ft.',H)
fprintf('\nThe surface area of the silo is: %f square ft.',S)
```

The Command Window where the script file, named silo, was executed is:

```
>> r = 30; R = 45; V = 200000;     Assigning values to r, R, and V.
>> silo                            Running the script file named silo.
```

The height H is: 64.727400 ft.

The surface area of the silo is: 15440.777753 square ft.

Sample Problem 4-2:   Centroid of a composite area

Write a program in a script file that calculates the coordinates of the centroid of a
composite area. (A composite area can
easily be divided into sections whose
centroids are known.) The user needs to
divide the area to sections and know the
coordinates of the centroid (two numbers) and the area of each section (one
number). When the script file is executed,
it asks the user to enter the three numbers
as a row in a matrix. The user enters as
many rows as there are sections. A section that represents a hole is taken to have
a negative area. For output, the program

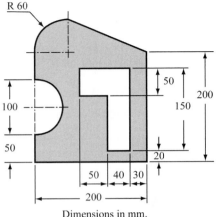

displays the coordinates of the centroid of the composite area. Use the program to
calculate the centroid of the area shown in the figure.

Solution

The area is divided into six sections as shown in the following figure. The total

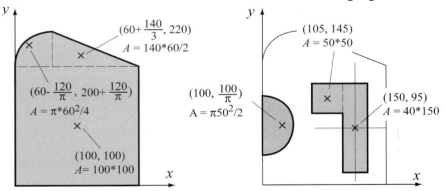

Units: coordinates mm, area mm².

area is calculated by adding the three sections on the left and subtracting the three
sections on the right. The location and coordinates of the centroid of each section
are marked in the figure, as well as the area of each section.

The coordinates $\bar{X}$ and $\bar{Y}$ of the centroid of the total area are given by:

$$\bar{X} = \frac{\Sigma A \bar{x}}{\Sigma A} \text{ and } \bar{Y} = \frac{\Sigma A \bar{y}}{\Sigma A}, \text{ where } \bar{x}, \bar{y}, \text{ and } A \text{ are the coordinates of the centroid}$$

and area of each section, respectively.

A script file with a program for calculating the coordinates of the centroid of a
composite area is:

The script file was saved with the name Centroid. The following shows the Com-

% The program calculates the coordinates of the centroid

% of a composite area.

clear C xs ys As

C=input('Enter a matrix in which each row has three elements.\nIn each row enter the x and y coordinates of the centroid and the area of a section.\n');

xs = C(:,1)';

> Creating a row vector for the x coordinate of each section (first column of C).

ys = C(:,2)';

> Creating a row vector for the y coordinate of each section (second column of C).

As = C(:,3)';

> Creating a row vector for the area of each section (third column of C).

A = sum(As);

> Calculating the total area.

x = sum(As.*xs)/A;

y = sum(As.*ys)/A;

> Calculating the coordinates of the centroid of the composite area.

fprintf('The coordinates of the centroid are: (%f, %f)',x,y)

mand Window where the script file was executed.

>> Centroid

Enter a matrix in which each row has three elements.

In each row enter the x and y coordinates of the centroid and the area of a section.

[100 100 200*200
60-120/pi 200+120/pi pi*60^2/4
60+140/3 220 140*60/2
100 100/pi -pi*50^2/2
150 95 -40*150
105 145 -50*50]

> Entering the data for matrix C. Each row has three elements: the x, y, and A of a section.

The coordinates of the centroid are: (85.387547 , 131.211809)
>>

Sample Problem 4-3: Voltage divider

When several resistors are connected in an electrical circuit in series, the voltage across each of them is given by the voltage divider rule:

$$v_n = \frac{R_n}{R_{eq}} v_s$$

where v_n and R_n are the voltage across resistor n and its resistance, respectively, $R_{eq} = \Sigma R_n$ is the equivalent resistance, and v_s is the source voltage. The power dissipated in each resistor is given by:

$$P_n = \frac{R_n}{R_{eq}^2} v_s^2$$

The figure below shows, for example, a circuit with seven resistors connected in series.

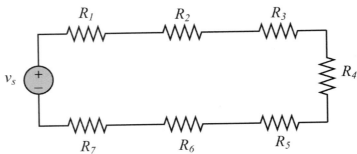

Write a program in a script file that calculates the voltage across each resistor, and the power dissipated in each resistor, in a circuit that has resistors connected in series. When the script file is executed it requests the user to first enter the source voltage and then to enter the resistances of the resistors in a vector. The program displays a table with the resistances listed in the first column, the voltage across the resistor in the second, and the power dissipated in the resistor in the third column. Following the table, the program displays the current in the circuit, and the total power.

Run the file and enter the following data for v_s and the R's.

$v_s = 24V$, $R_1 = 20\Omega$, $R_2 = 14\Omega$, $R_3 = 12\Omega$, $R_4 = 18\Omega$, $R_5 = 8\Omega$, $R_6 = 15\Omega$, $R_7 = 10\Omega$.

Solution

A script file that solves the problem is shown below.

```
% The program calculates the voltage across each resistor
% in a circuit that has resistors connected in series.
vs = input('Please enter the source voltage ');
Rn = input('Enter the values of the resistors as elements in a row vector\n');
```

```
Req = sum(Rn);
vn = Rn*vs/Req;
Pn = Rn*vs^2/Req^2;
i = vs/Req;
Ptotal = vs*i;
Table = [Rn', vn', Pn'];
disp(' ')
disp(' Resistance Voltage   Power')
disp('  (Ohms)   (Volts) (Watts)')
disp(' ')
disp(Table)
disp(' ')
fprintf('The current in the circuit is %f Amps.',i)
fprintf('\nThe total power dissipated in the circuit is %f Watts.',Ptotal)
```

`Req = sum(Rn);`	Calculate the equivalent resistance.
`vn = Rn*vs/Req;`	Apply the voltage divider rule.
`Pn = Rn*vs^2/Req^2;`	Calculate the power in each resistor.
`i = vs/Req;`	Calculate the current in the circuit.
`Ptotal = vs*i;`	Calculate the total power in the circuit.
`Table = [Rn', vn', Pn'];`	Create a variable `table` with the vectors Rn, vn, and Pn as columns.
`disp(' Resistance Voltage Power')`	Display headings for the columns.
`disp(' ')`	Display an empty line.
`disp(Table)`	Display the variable `Table`.

The Command Window where the script file was executed is:

```
>> VoltageDivider
Please enter the source voltage 24
Enter the value of the resistors as elements in a row vector
[20  14  12  18  8  15  10]
```

Name of the script file.

Source voltage entered by the user.

Resistor values entered as a vector.

Resistance	Voltage	Power
(Ohms)	(Volts)	(Watts)
20.0000	4.9485	1.2244
14.0000	3.4639	0.8571
12.0000	2.9691	0.7346
18.0000	4.4536	1.1019
8.0000	1.9794	0.4897
15.0000	3.7113	0.9183
10.0000	2.4742	0.6122

The current in the circuit is 0.247423 Amps.

The total power dissipated in the circuit is 5.938144 Watts.

4.9 PROBLEMS

Note: *In addition to the problems below, the problems at the end of Chapter 3 can also be solved by writing script files.*

Solve the following problems by first writing a program in a script file and then running the file in the Command Window.

1. A cone shaped paper cup is designed to have a volume of 250 cm^3. Determine the radius, r, of the base and the surface area, S, of the paper for cups with heights, h of 5, 6, 7, 8, and 9 cm.
    The volume, V, and the surface area of the paper are given by:

    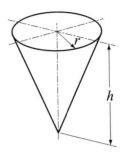

    $$V = \frac{1}{3}\pi r^2 h \qquad S = \pi r \sqrt{r^2 + h^2}$$

2. In a movie theater the angle θ at which a viewer sees the picture on the screen depends on the distance x of the viewer from the screen. For a movie theater with the dimensions shown in the figure, determine the angle θ (in degrees) for viewers sitting at distances of 30, 45, 60, 75, and 90 ft. from the screen.

    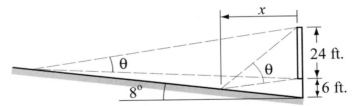

3. The population of a certain country is 50 million and is expected to double in 20 years. Calculate the population 5, 10, and 15 years from now by defining a vector t with 3 elements and using element-by-element calculations. Population growth can be modeled by the equation $P = P_0 2^{t/d}$, where P is the population at time t, P_0 is the population at $t = 0$, and d is the doubling time.

4. A hiker needs to cross a sandy area in order to get from point A to a camp site at point B. He can do so by crossing the sand perpendicular to the trail and then walking along the trail, or by crossing the sand in an angle θ up to the trail, and then walking along the trail. The hiker walks 3.5 km/h in the sand and 5 km/h on the trail. Determine the time it will take him to reach the camp

site for $\theta = 0, 10, 20, 30, 40, 50$ and 60 degrees.

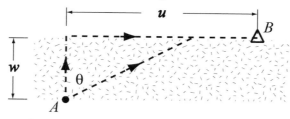

The distances w and u are $w = 4.5$ km, and $u = 14$ km. Write a program in a script file to solve the problem. Define all the variables within the script file. Display the results in a two column table in which the first column is θ and the second column is the corresponding t.

5. Write a script file that determines the balance in a saving account at the end of every year for the first 10 years. The account has an initial investment of $1,000 and interest rate of 6.5% that compounds annually. Display the information in a table.

    For an initial investment of A, and interest rate of r, the balance B after n years is given by:

    $$B = A\left(1 + \frac{r}{100}\right)^n$$

6. The velocity, v, and the distance, d, as a function of time, of a car that accelerates from rest at constant acceleration, a, are given by:

    $$v(t) = at \quad \text{and} \quad d(t) = \frac{1}{2}at^2$$

    Determine v and d as every second for the first 10 seconds for a car with acceleration of $a = 1.55$ m/s². Display the results in a three-column table in which the first column is time (s), the second distance (m), and the third is velocity (m/s).

7. When several resistors are connected in an electrical circuit in parallel, the current through each of them is given by:

    $$i_n = \frac{v_s}{R_n}$$

    where i_n and R_n are the current through resistor n and its resistance, respectively, and v_s is the source voltage. The equivalent resistance, R_{eq}, can be determine from the equation:

    $$\frac{1}{R_{eq}} = \frac{1}{R_1} + \frac{1}{R_2} + \dots + \frac{1}{R_n}$$

    The source current is given by: $i_s = v_s/R_{eq}$, and the power, P_n, dissipated in each resistor is given by: $P_n = v_s i_n$.

Write a program in a script file that calculates the current through each resistor and the power dissipated in each in a circuit that has resistors connected in parallel. When the script file runs it asks the user first to enter the source voltage and then to enter the resistors' resistance in a vector. The program displays a table with the resistances listed in the first column, the current through the resistor in the second, and the power dissipated in the resistor in the third column. Following the table, the program displays the source current and the total power.

Use the script file for the circuit shown below.

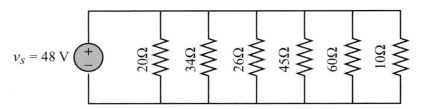

8. The graph of the function $f(x) = ax^3 + bx^2 + cx + d$ passes through the points $(-2, -3.4)$, $(-0.5, 5.525)$, $(1, 16.7)$, and $(2.5, 70.625)$. Determine the constants a, b, c, and d. (Write a system of four equations with four unknowns and use MATLAB to solve the equations.)

9. A truss is a structure made of members jointed at their ends. For the truss shown in the figure, the forces in the seven members are determined by solving the following set of seven equations.

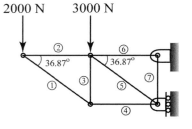

$F_1 \sin(36.87°) = -2000$

$F_1 \cos(36.87°) + F_2 = 0$

$F_3 + F_1 \sin(36.87°) = 0$

$F_4 - F_1 \cos(36.87°) = 0$

$-F_3 - F_5 \sin(36.87°) = 3000$

$F_6 + F_5 \cos(36.87°) - F_2 = 0$

$F_5 \sin(36.87°) + F_7 = 0$

Write the equations in a matrix form and use MATLAB to determine the forces in the members. A positive force means tensile force and a negative force means compressive force. Display the results in a table.

Chapter 5
Two-Dimensional Plots

Plots are a very useful tool for presenting information. This is true in any field, but especially in science and engineering where MATLAB is mostly used. MATLAB has many commands that can be used for creating different types of plots. These include standard plots with linear axes, plots with logarithmic and semi-logarithmic axes, bar and stairs plots, polar plots, three-dimensional contour surface and mesh plots, and many more. The plots can be formatted to have a desired appearance. The line type (solid, dashed, etc.), color, and thickness can be prescribed, line markers and grid lines can be added, as well as titles and text comments. Several graphs can be plotted in the same plot and several plots can be placed on the same page. When a plot contains several graphs and/or data points, a legend can be added to the plot as well.

This chapter describes how MATLAB can be used to create and format many types of two-dimensional plots. Three-dimensional plots are addressed separately in Chapter 9. An example of a simple two-dimensional plot that was created with MATLAB is shown in Figure 5-1. The figure contains two curves that show the variation of light intensity with distance. One curve is constructed from data points measured in an experiment, and the other curve shows the variation of light as predicted by a theoretical model. The axes in the figure are both linear, and different types of lines (one solid and one dashed) are used for the curves. The theoretical curve is shown with a solid line, while the experimental points are connected with a dashed line. Each data point is marked with a circular marker. The dashed line that connects the experimental points is actually red when the plot is displayed in the Figure Window. As shown, the plot in Figure 5-1 is formatted to have a title, axes' titles, a legend, markers, and a boxed text label.

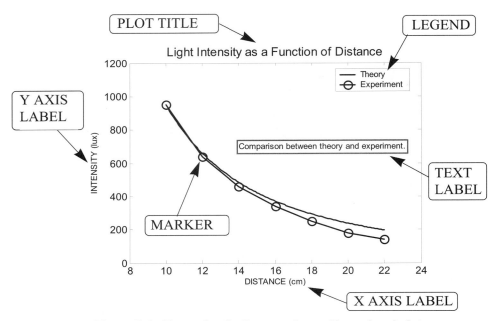

Figure 5-1: Example of a formatted two-dimensional plot.

5.1 THE plot *COMMAND*

The plot command is used to create two-dimensional plots. The simplest form of the command is:

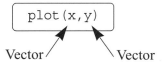

The arguments x and y are each a vector (one-dimensional array). Both vectors **must** have the same number of elements. When the plot command is executed a figure is created in the Figure Window. If not already open, the Figure Window opens automatically when the command is executed. The figure has a single curve with the x values on the abscissa (horizontal axis), and the y values on the ordinate (vertical axis). The curve is constructed of straight-line segments that connect the points whose coordinates are defined by the elements of the vectors x and y. The vectors, of course, can have any name. The vector that is typed first in the plot command is used for the horizontal axis, and the vector that is typed second is used for the vertical axis.

    The figure that is created has axes with linear scale and default range. For example, if a vector *x* has the elements 1, 2, 3, 5, 7, 7.5, 8, 10, and a vector *y* has the elements 2, 6.5, 7, 7, 5.5, 4, 6, 8, a simple plot of *y* versus *x* can be created by typing the following in the Command Window:

```
>> x = [1  2  3  5  7  7.5  8  10];
>> y = [2  6.5  7  7  5.5  4  6  8];
>> plot(x,y)
```

Once the plot command is executed, the Figure Window opens and the plot is displayed, as shown in Figure 5-2.

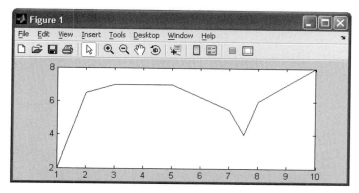

Figure 5-2: The Figure Window with a simple plot.

The plot appears on the screen in blue which is the default line color.

The plot command has additional optional arguments that can be used to specify the color and style of the line and the color and type of markers, if any are desired. With these options the command has the form:

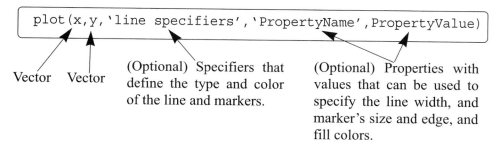

Vector   Vector   (Optional) Specifiers that define the type and color of the line and markers.   (Optional) Properties with values that can be used to specify the line width, and marker's size and edge, and fill colors.

Line Specifiers:

Line specifiers are optional and can be used to define the style and color of the line and the type of markers (if markers are desired). The line style specifiers are:

Line Style	Specifier		Line Style	Specifier
solid (default)	-		dotted	:
dashed	--		dash-dot	-.

The line color specifiers are:

Line Color	Specifier
red	r
green	g
blue	b
cyan	c

Line Color	Specifier
magenta	m
yellow	y
black	k
white	w

The marker type specifiers are:

Marker Type	Specifier		Marker Type	Specifier
plus sign	+		square	s
circle	o		diamond	d
asterisk	*		five-pointed star	p
point	.		six-pointed star	h

Notes about using the specifiers:

- The specifiers are typed inside the `plot` command as strings.
- Within the string the specifiers can be typed in any order.
- The specifiers are optional. This means that none, one, two, or all the three can be included in a command.

Some examples:

`plot(x,y)`	A blue solid line connects the points with no markers (default).
`plot(x,y,'r')`	A red solid line connects the points.
`plot(x,y,'--y')`	A yellow dashed line connects the points.
`plot(x,y,'*')`	The points are marked with * (no line between the points).
`plot(x,y,'g:d')`	A green dotted line connects the points that are marked with diamond markers.

Property Name and Property Value:

Properties are optional and can be used to specify the thickness of the line, the size of the marker, and the colors of the marker's edge line and fill. The Property Name is typed as a string, followed by a comma and a value for the property, all inside the `plot` command.

Four properties and possible values are:

Property Name	Description	Possible Property Values
`LineWidth` (or `linewidth`)	Specifies the width of the line.	A number in units of points (default 0.5).
`MarkerSize` (or `markersize`)	Specifies the size of the marker.	A number in units of points.
`MarkerEdge-Color` (or `markeredge-color`)	Specifies the color of the marker, or the color of the edge line for filled markers.	Color specifiers from the table above, typed as a string.
`MarkerFace-Color` (or `markerface-color`)	Specifies the color of the filling for filled markers.	Color specifiers from the table above, typed as a string.

For example, the command:

```
plot(x,y,'-mo','LineWidth',2,'markersize',12,
        'MarkerEdgeColor','g','markerfacecolor','y')
```

creates a plot that connects the points with a magenta solid line and circles as markers at the points. The line width is two points and the size of the circle markers is 12 points. The markers have a green edge line and yellow filling.

A note about line specifiers and properties:

The three line specifiers, which are the style and color of the line, and the type of the marker can also be assigned with a `PropertyName` argument followed by a `PropertyValue` argument. The Property Names for the line specifiers are:

Specifier	Property Name	Possible Property Values
Line Style	`linestyle` (or `LineStyle`)	Line style specifier from the table above, typed as a string.
Line Color	`color` (or `Color`)	Color specifiers from the table above, typed as a string.

Specifier	Property Name	Possible Property Values
Marker	`marker` (or `Marker`)	Marker specifier from the table above, typed as a string.

As with any command, the `plot` command can be typed in the Command Window, or it an be included in a script file. It also can be used in a function file (explained in Chapter 6). It should also be remembered that before the `plot` command can be executed the vectors x and y must have assigned elements. This can be done, as was explained in Chapter 2, by entering values directly, by using commands, or as the result of mathematical operations. The next two subsections show examples of creating simple plots.

5.1.1 Plot of Given Data

In this case given data is first used to create vectors that are then used in the `plot` command. For example, the following table contains sales data of a company from 1988 to 1994.

YEAR	1988	1989	1990	1991	1992	1993	1994
SALES (millions)	8	12	20	22	18	24	27

To plot this data, the list of years is assigned to one vector (named `yr`), and the corresponding sale data is assigned to a second vector (named `sle`). The Command Window where the vectors are created and the `plot` command is used is shown below:

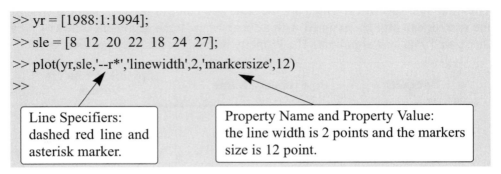

Once the `plot` command is executed the Figure Window with the plot, as shown in Figure 5-3, opens. The plot appears on the screen in red.

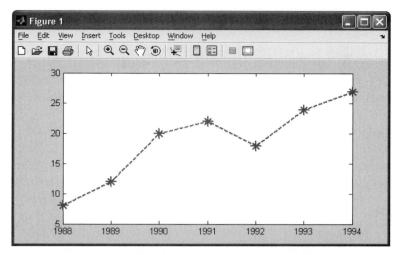

Figure 5-3: The Figure Window with a plot of the sales data.

5.1.2 Plot of a Function

In many situations there is a need to plot a given function. This can be done in MATLAB by using the `plot` or the `fplot` commands. The use of the `plot` command is explained below. The `fplot` command is explained in detail in the next section.

In order to plot a function $y = f(x)$ with the `plot` command, the user needs to first create a vector of values of x for the domain that the function will be plotted. Then, a vector y is created with the corresponding values of $f(x)$ by using element-by-element calculations (see Chapter 3). Once the two vectors exist, they can be used in the `plot` command.

As an example, the `plot` command is used to plot the function $y = 3.5^{-0.5x}\cos(6x)$ for $-2 \le x \le 4$. A program that plots this function is shown in the following script file.

% A script file that creates a plot of	
% the function: 3.5.^(-0.5*x).*cos(6x)	
x = [-2:0.01:4];	Create vector x with the domain of the function.
y = 3.5.^(-0.5*x).*cos(6*x);	Create vector y with the function value at each x.
plot(x,y)	Plot y as a function of x.

Once the script file is executed, the plot is created in the Figure Window, as shown in Figure 5-4.

Since the plot is made up of segments of straight lines that connect the points, to obtain an accurate plot of a function, the spacing between the elements of the vector x must be appropriate. Smaller spacing is needed for a function that

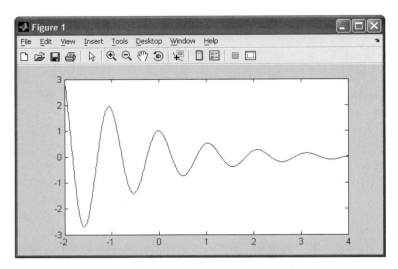

Figure 5-4: **The Figure Window with a plot of the function:** $y = 3.5^{-0.5x}\cos(6x)$.

changes rapidly. In the last example a small spacing of 0.01 produced the plot that is shown in Figure 5-4. However, if the same function in the same domain is plotted with much larger spacing, for example 0.3, the plot that is obtained, shown in Figure 5-5, gives a distorted picture of the function. Note also that in Figure 5-4,

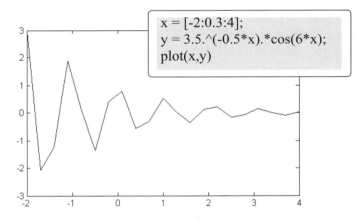

```
x = [-2:0.3:4];
y = 3.5.^(-0.5*x).*cos(6*x);
plot(x,y)
```

Figure 5-5: A plot of the function $y = 3.5^{-0.5x}\cos(6x)$ **with large spacing.**

the plot is shown with the Figure Window, while in Figure 5-5, only the plot is shown. The plot can be copied from the Figure Window (in the **Edit** menu select **Copy Figure**) and then pasted in other applications.

5.2 *THE* fplot *COMMAND*

The fplot command plots a function with the form $y = f(x)$ between specified limits. The command has the form:

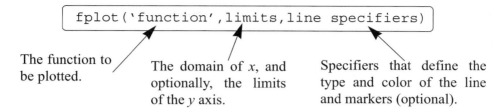

The function to be plotted.

The domain of x, and optionally, the limits of the y axis.

Specifiers that define the type and color of the line and markers (optional).

'function': The function can be typed directly as a string inside the command. For example, if the function that is being plotted is $f(x) = 8x^2 + 5\cos(x)$, it is typed as: `'8*x^2+5*cos(x)'`. The function can include MATLAB built-in functions and functions that are created by the user (covered in Chapter 6).

- The function to be plotted can be typed as a function of any letter. For example, the function in the previous paragraph can be typed as: `'8*z^2+5*cos(z)'`, or `'8*t^2+5*cos(t)'`.

- The function can not include previously defined variables. For example, in the function above it is not possible to assign 8 to a variable, and then use the variable when the function is typed in the `fplot` command.

limits: The limits is a vector with two elements that specify the domain of x [`xmin,xmax`], or a vector with four elements that specifies the domain of x and the limits of the y-axis [`xmin,xmax,ymin,ymax`].

Line specifiers: The line specifiers are the same as in the plot command. For example, a plot of the function $y = x^2 + 4\sin(2x) - 1$ for $-3 \le x \le 3$ can be created with the `fplot` command by typing:

```
>> fplot('x^2 + 4*sin(2*x) - 1',[-3 3])
```

in the Command Window. The figure that is obtained in the Figure Window is shown in Figure 5-6.

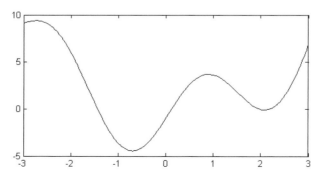

Figure 5-6: A plot of the function $y = x^2 + 4\sin(2x) - 1$.

5.3 PLOTTING MULTIPLE GRAPHS IN THE SAME PLOT

In many situations there is a need to make several graphs in the same plot. This is shown, for example, in Figure 5-1 where two graphs are plotted in the same figure. There are three methods to plot multiple graphs in one figure. One is by using the `plot` command, the other is by using the `hold on`, `hold off` commands, and the third is by using the `line` command.

5.3.1 Using the `plot` Command

Two or more graphs can be created in the same plot by typing pairs of vectors inside the `plot` command. The command:

$$\texttt{plot(x,y,u,v,t,h)}$$

creates three graphs: y vs. x, v vs. u, and h vs. t, all in the same plot. The vectors of each pair must be of the same length. MATLAB automatically plots the graphs in different colors so that they can be identified. It is also possible to add line specifiers following each pair. For example the command:

$$\texttt{plot(x,y,`-b',u,v,`--r',t,h,`g:')}$$

plots y vs. x with a solid blue line, v vs.u with a dashed red line, and h vs. t with a dotted green line.

Sample Problem 5-1:   Plotting a function and its derivatives

Plot the function $y = 3x^3 - 26x + 10$, and its first and second derivatives, for $-2 \leq x \leq 4$, all in the same plot.

Solution

The first derivative of the function is: $y' = 9x^2 - 26$.
The second derivative of the function is: $y'' = 18x$.
A script file that creates a vector x, and calculates the values of y, y', and y'' is:

```
x = [-2:0.01:4];                Create vector x with the domain of the function.
y = 3*x.^3 - 26*x + 6;          Create vector y with the function value at each x.
yd = 9*x.^2 - 26;               Create vector yd with values of the first derivative.
ydd = 18*x;                     Create vector ydd with values of the second derivative.
plot(x,y,'-b',x,yd,'--r',x,ydd,':k')
                                Create three graphs, y vs. x, yd vs. x,
                                and ydd vs. x in the same figure.
```

The plot that is created is shown in Figure 5-7.

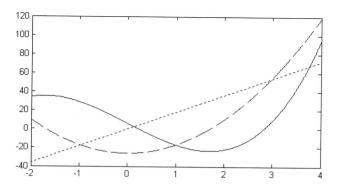

Figure 5-7: A plot of the function $y = 3x^3 - 26x + 10$ **and its first and second derivatives.**

5.3.2 *Using the* hold on, hold off *Commands*

To plot several graphs using the hold on, hold off commands, one graph is plotted first with the plot command. Then the hold on command is typed. This keeps the Figure Window with the first plot open, including the axis properties and formatting (see Section 5.4) if any was done. Additional graphs can be added with plot commands that are typed next. Each plot command creates a graph that is added to that figure. The hold off command stops this process. It returns MATLAB to the default mode in which the plot command erases the previous plot and resets the axis properties.

As an example, a solution of Sample Problem 5-1 using the hold on, hold off commands, is shown in the following script file:

```
x = [-2:0.01:4];
y = 3*x.^3 - 26*x + 6;
yd = 9*x.^2 - 26;
ydd = 18*x;
plot(x,y,'-b')          The first graph is created.
hold on
plot(x,yd,'--r')        Two more graphs are added to the figure.
plot(x,ydd,':k')
hold off
```

5.3.3 *Using the* line *Command*

With the line command additional graphs (lines) can be added to a plot that already exists. The form of the line command is:

(Optional) Properties with values that can be
used to specify the line style, color, and width,
marker type, size, and edge and fill colors.

The format of the `line` command is almost the same as the `plot` command (see
Section 5.1). The `line` command does not have the line specifiers, but the line
style, color, and marker can be specified with the Property Name and property
value features. The properties are optional and if none are entered MATLAB uses
default properties and values. For example, the command:

```
line(x,y,'linestyle','--','color','r','marker','o')
```

will add a dashed red line with circular markers to a plot that already exists.

The major difference between the `plot` and `line` commands is that the
`plot` command starts a new plot every time it is executed, while the `line` command adds lines to a plot that already exists. To make a plot that has several
graphs, a plot command is typed first and then line commands are typed for additional graphs. (If a line command is entered before a plot command an error message is displayed.)

The solution to Sample Problem 5-1, which is the plot in Figure 5-7, can be
obtained by using the `plot` and `line` commands as shown in the following
script file:

```
x = [-2:0.01:4];
y = 3*x.^3 - 26*x + 6;
yd = 9*x.^2 - 26;
ydd = 18*x;
plot(x,y,'LineStyle','-','color','b')
line(x,yd,'LineStyle','--','color','r')
line(x,ydd,'linestyle',':','color','k')
```

5.4 FORMATTING A PLOT

The `plot` and `fplot` commands create bare plots. Usually, however, a figure
that contains a plot needs to be formatted to have a specific look and to display
information in addition to the graph itself. It can include specifying axis labels,
plot title, legend, grid, range of custom axis, and text labels.

Plots can be formatted by using MATLAB commands that follow the `plot`
or `fplot` commands, or interactively by using the plot editor in the Figure Win-

dow. The first method is useful when a `plot` command is a part of a computer program (script file). When the formatting commands are included in the program, a formatted plot is created every time the program is executed. On the other hand, formatting that is done in the Figure Window with the plot editor after a plot has been created holds only for that specific plot, and will have to be repeated the next time the plot is created.

5.4.1 Formatting a Plot Using Commands

The formatting commands are entered after the `plot` or the `fplot` commands. The various formatting commands are:

The `xlabel` and `ylabel` commands:

Labels can be placed next to the axes with the `xlabel` and `ylabel` commands which have the form:

```
xlabel('text as string')
ylabel('text as string')
```

The `title` command:

A title can be added to the plot with the command:

```
title('text as string')
```

The text is placed at the top of the figure as a title.

The `text` command:

A text label can be placed in the plot with the `text` or `gtext` commands:

```
text(x,y,'text as string')
gtext('text as string')
```

The `text` command places the text in the figure such that the first character is positioned at the point with the coordinates x, y (according to the axes of the figure). The `gtext` command places the text at a position specified by the user. When the command is executed, the Figure Window opens and the user specifies the position with the mouse.

The `legend` command:

The `legend` command places a legend on the plot. The legend shows a sample of the line type of each graph that is plotted, and places a label, specified by the user, beside the line sample. The form of the command is:

```
legend('string1','string2', ..... ,pos)
```

The strings are the labels that are placed next to the line sample. Their order corresponds to the order that the graphs were created. The `pos` is an optional number that specifies where in the figure the legend is placed. The options are:

`pos = -1`	Places the legend outside the axes boundaries on the right side.
`pos = 0`	Places the legend inside the axes boundaries in a location that interferes the least with the graphs.
`pos = 1`	Places the legend at the upper-right corner of the plot (default).
`pos = 2`	Places the legend at the upper-left corner of the plot.
`pos = 3`	Places the legend at the lower-left corner of the plot.
`pos = 4`	Places the legend at the lower-right corner of the plot.

Formatting the text in the `xlabel`, `ylabel`, `title`, `text`

and `legend` **commands:**

The text in the string that is included in the commands and is displayed when the commands are executed can be formatted. The formatting can be used to define the font, size, position (superscript, subscript), style (italic, bold, etc.), and color of the characters, color of the background, and to define many other details of the display. Some of the more common formatting possibilities are described below. A complete explanation of all the formatting features can be found in the Help Window under Text and Text Properties. The formatting can be done either by adding modifiers inside the string, or by adding to the command optional `PropertyName` and `PropertyValue` arguments following the string.

The modifiers are characters that are inserted within the string. Some of the modifiers that can be added are:

Modifier	Effect	Modifier	Effect
`\bf`	bold font.	`\fontname{fontname}`	specified font is used.
`\it`	italic style.	`\fontsize{fontsize}`	specified font size is used.
`\rm`	normal font.		

These modifiers affect the text from the point that they are inserted until the end of the string. It is also possible to have the modifiers applied to only a section of the string by typing the modifier and the text to be affected inside braces { }.

Subscript and superscript:

A single character can be displayed as a subscript or a superscript by typing _ (the underscore character) or ^ in front of the character, respectively. Several consecutive characters can be displayed as subscript or a superscript by typing the characters inside braces { } following the _ or the ^.

Greek characters:

Greek characters can be included in the text by typing \name of the let-ter within the string. To display a lowercase Greek letter the name of the letter should be typed in all lowercase English characters, To display a capital Greek let-ter the name of the letter should start with a capital letter. Some examples are:

Characters in the string	Greek Letter
\alpha	α
\beta	β
\gamma	γ
\theta	θ
\pi	π
\sigma	σ

Characters in the string	Greek Letter
\Phi	Φ
\Delta	Δ
\Gamma	Γ
\Lambda	Λ
\Omega	Ω
\Sigma	Σ

Formatting of the text that is displayed by the xlabel, ylabel, title, and text commands can also be done by adding optional PropertyName and PropertyValue arguments following the string inside the command. With this option the text command, for example, has the form:

```
text(x,y,'text as string',PropertyName,PropertyValue)
```

In the other three commands the PropertyName and PropertyValue argu-ments are added in the same way. The PropertyName is typed as a string, and the PropertyValue is typed as a number if the property value is a number and as a string if the property value is a word or a letter character. Some of the Prop-erty Names and corresponding possible Property Values are:

Property Name	Description	Possible Property Values
Rotation	Specifies the orientation of the text.	Scalar (degrees) Default: 0
FontAngle	Specifies italic or normal style characters.	normal, italic Default: normal
FontName	Specifies the font for the text.	Font name that is available in the system.
FontSize	Specifies the size of the font.	Scalar (points) Default: 10

Property Name	Description	Possible Property Values
`FontWeight`	Specifies the weight of the characters.	`light`, `normal`, `bold` Default: `normal`
`Color`	Specifies the color of the text.	Color specifiers (See Section 5.1).
`Background-Color`	Specifies the background color (rectangular area).	Color specifiers (See Section 5.1).
`EdgeColor`	Specifies the color of the edge of a rectangular box around the text.	Color specifiers (See Section 5.1). Default: none.
`LineWidth`	Specifies the width of the edge of a rectangular box around the text.	Scalar (points) Default: 0.5

The `axis` command:

When the `plot (x,y)` command is executed, MATLAB creates axes with limits that are based on the minimum and maximum values of the elements of x and y. The `axis` command can be used to change the range and the appearance of the axes. In many situations a graph looks better if the range of the axes extend beyond the range of the data. The following are some of the possible forms of the axis command:

`axis([xmin,xmax,ymin,ymax])`   Sets the limits of both the x and y axes (xmin, xmax, ymin, and ymax are numbers).

`axis equal`   Sets the same scale for both axes.

`axis square`   Sets the axes region to be square.

`axis tight`   Sets the axis limits to the range of the data.

The `grid` command:

`grid on`   Adds grid lines to the plot.

`grid off`   Removes grid lines from the plot.

An example of formatting a plot by using commands is given in the following script file that was used to generate the formatted plot in Figure 5-1.

```
x = [10:0.1:22];
y = 95000./x.^2;
```

```
xd = [10:2:22];
yd = [950  640  460  340  250  180  140];
plot(x,y,'-','LineWidth',1.0)
xlabel('DISTANCE (cm)')
ylabel('INTENSITY (lux)')
title('\fontname{Arial}Light Intensity as a Function of Distance','FontSize',14)
axis([8 24 0 1200])
text(14,700,'Comparison between theory and experiment.','EdgeColor','r','LineWidth',2)
hold on
plot(xd,yd,'ro--','linewidth',1.0,'markersize',10)
legend('Theory','Experiment',0)
hold off
```

> Formatting text inside the `title` command.

> Formatting text inside the `text` command.

5.4.2 Formatting a Plot Using the Plot Editor

A plot can be formatted interactively in the Figure Window by clicking on the plot and/or using the menus. Figure 5-8 shows the Figure Window with the plot of Figure 5-1. The Plot Editor can be used to introduce new formatting items, or to

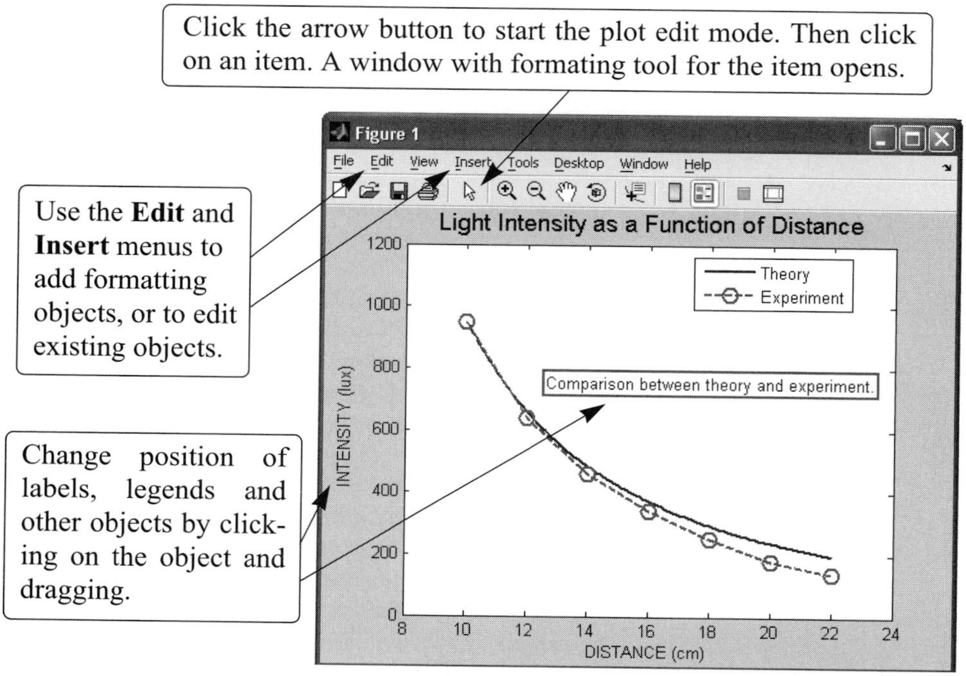

> Click the arrow button to start the plot edit mode. Then click on an item. A window with formating tool for the item opens.

> Use the **Edit** and **Insert** menus to add formatting objects, or to edit existing objects.

> Change position of labels, legends and other objects by clicking on the object and dragging.

Figure 5-8: Formatting a plot using the plot editor.

modify formatting that was initially introduced with the formatting commands.

5.5 PLOTS WITH LOGARITHMIC AXES

Many science and engineering applications require plots in which one or both axes have a logarithmic (log) scale. Log scales provide means for presenting data over a wide range of values. It also provides a tool for identifying characteristics of data and possible forms of mathematical relationships that can be appropriate for modeling the data (see Section 8.2.2).

MATLAB commands for making plots with log axes are:

`semilogy(x,y)`   Plots y versus x with a log (base 10) scale for the y axis and linear scale for the x axis.

`semilogx(x,y)`   Plots y versus x with a log (base 10) scale for the x axis and linear scale for the y axis.

`loglog(x,y)`   Plots y versus x with a log (base 10) scale for both axes.

Line specifiers and Property Name and property value can be added to the commands (optional) just as in the `plot` command. As an example, Figure 5-9 shows

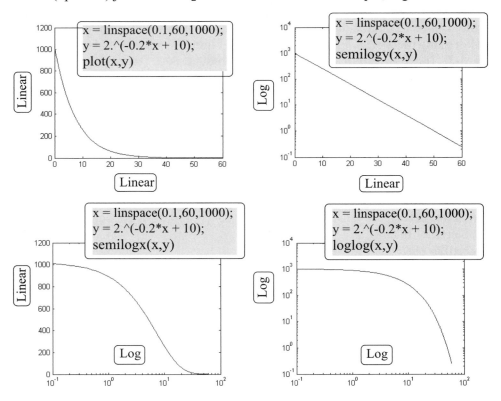

Figure 5-9: Plots of $y = 2^{(-0.2x + 10)}$ with linear, semilog, and loglog scales.

a plot of the function $y = 2^{(-0.2x+10)}$ for $0.1 \le x \le 60$. The figure shows four plots of the same function: one with linear axes, one with log scale for the y-axis, one with log scale for the x-axis, and one with log scale on both axes.

Notes for plots with logarithmic axes:

- The number zero can not be plotted on a log scale (since a log of zero is not defined).

- Negative numbers can not be plotted on log scales (since a log of a negative number is not defined).

5.6 PLOTS WITH SPECIAL GRAPHICS

All the plots that have been presented so far in this chapter are line plots in which the data points are connected by lines. In many situations plots with different graphics or geometry can present data more effectively. MATLAB has many options for creating a wide variety of plots. These include bar, stairs, stem, pie, plots and many more. The following shows some of the special graphics plots that can be created with MATLAB. A complete list of the plotting functions that MAT-LAB has and information of how to use them can be found in Help Window. In this window first choose "Functions by Category,", then select "Graphics" and then select "Basic Plots and Graphs" or "Specialized Plotting."

Bar (vertical and horizontal), stairs, and stem plots are presented below using the sales data from Section 5.1.1.

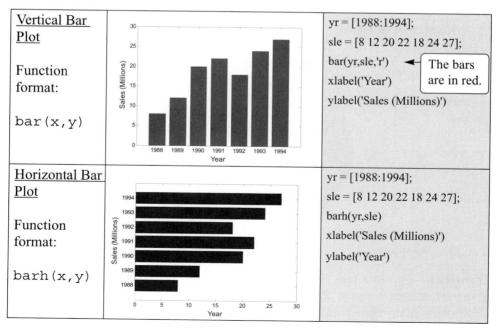

Stairs Plot
 	
Function format:
 	
`stairs(x,y)`
	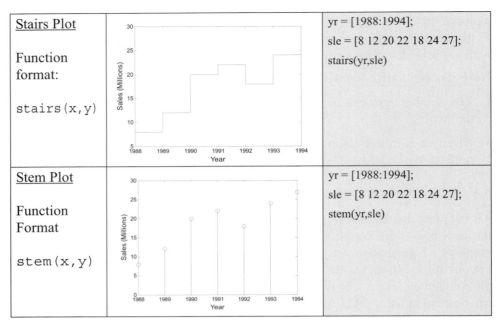	yr = [1988:1994];
		
		sle = [8 12 20 22 18 24 27];
		
		stairs(yr,sle)

Stem Plot
 	
Function Format
 	
`stem(x,y)`
		yr = [1988:1994];
		
		sle = [8 12 20 22 18 24 27];
		
		stem(yr,sle)

Pie charts are useful for visualizing the relative sizes of different but related quantities. For example, the table below shows the grades that were assigned to a class. The data below is used to create the pie chart that follows.

Grade	A	B	C	D	E
Number of Students	11	18	26	9	5

Pie Plot Function format: `pie(x)`	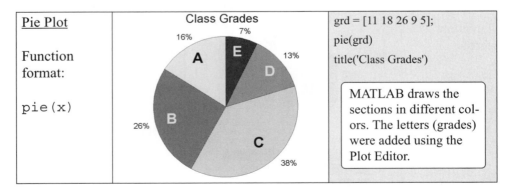	grd = [11 18 26 9 5]; pie(grd) title('Class Grades') MATLAB draws the sections in different colors. The letters (grades) were added using the Plot Editor.

5.7 HISTOGRAMS

Histograms are plots that show the distribution of data. The overall range of a given set of data points is divided to smaller subranges (bins), and the histogram shows how many data points are in each bin. The histogram is a vertical bar plot in which the width of each bar is equal to the range of the corresponding bin, and the height of the bar corresponds to the number of data points in the bin. Histograms are created in MATLAB with the `hist` command. The simplest form of the command is:

$$\boxed{\texttt{hist(y)}}$$

y   is a vector with the data points. MATLAB divides the range of the data points into 10 equally spaced subranges (bins), and then plots the number of data points in each bin.

    For example, the following data points are the daily maximum temperature (in °F) in Washington DC during the month of April, 2002: 58 73 73 53 50 48 56 73 73 66 69 63 74 82 84 91 93 89 91 80 59 69 56 64 63 66 64 74 63 69, (data from the U.S. National Oceanic and Atmospheric Administration). A histogram of this data is obtained with the commands:

```
>> y = [58  73  73  53  50  48  56  73  73  66  69  63  74  82  84  91  93  89  91  80
59  69  56  64  63  66  64  74  63  69];
>> hist(y)
```

The plot that is generated is shown in Figure 5-10 (the axes titles were added using the Plot Editor). The smallest value in the data set is 48 and the largest is 93,

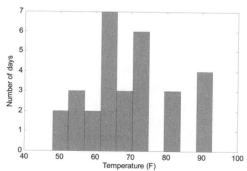

Figure 5-10: Histogram of temperature data.

which means that the range is 45 and that the width of each bin is 4.5. The range of the first bin is from 48 to 52.5 and contains two points. The range of the second bin is from 52.5 to 57 and contains three points, and so on. Two of the bins (75 to 79.5 and 84 to 88.5) do not contain any points.

    Since the division of the data range into 10 equally spaced bins might not be the division that is preferred by the user, the number of bins can be defined by the user to be different than 10. This can be done either by specifying the number of bins, or by specifying the center point of each bin as shown in the following two forms of the hist command:

| hist(y,nbins) | or | hist(y,x) |

nbins   is a scalar that defines the number of bins. MATLAB divides the range
         to equally spaced sub-ranges.

x   is a vector that specifies the location of the center of each bin (the dis-
         tance between the centers does not have to be the same for all the bins).
         The edges of the bins are at the middle point between the centers.

In the example above the user
might prefer to divide the temperature
range into 3 bins. This can be done with
the command:

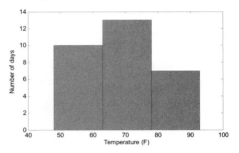

>> hist(y,3)

As shown on the right, the histogram
that is generated has three equally
spaced bins.

The number and width of the bins
can also be specified by a vector x
whose elements define the centers of
the bins. For example, shown on the
right is a histogram that displays the
temperature data from above in 6 bins
with an equal width of 10 degrees. The
elements of the vector x for this plot are
45, 55, 65, 75, 85, and 95. The plot was
obtained with the following commands:

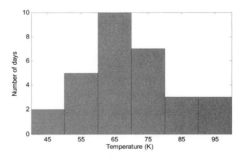

>> x = [45:10:95]

x =
    45   55   65   75   85   95
>> hist(y,x)

The hist command can be used with options that provide numerical output in addition to plotting a histogram. An output of the number of data points in
each bin can be obtained with one of the following commands:

| n=hist(y) | | n=hist(y,nbins) | | n=hist(y,x) |

The output n is a vector. The number of elements in n is equal to the number of
bins and the value of each element of n is the number of data points (frequency
count) in the corresponding bin. For example, the histogram in Figure 5-10 can

also be created with the following command:

>> n = hist(y)

n =

    2   3   2   7   3   6   0   3   0   4

> The vector n shows how may elements are in each bin.

The vector n shows that the first bin has 2 data points, the second bin has 3 data points, and so on.

An additional optional numerical output is the location of the bins. This output can be obtained with one of the following commands:

```
[n xout]=hist(y)
```

```
[n xout]=hist(y,nbins)
```

xout is a vector in which the value of each element is the location of the center of the corresponding bin. For example, for the histogram in Figure 5-10:

>> [n xout] = hist(y)

n =

    2   3   2   7   3   6   0   3   0   4

xout =

 50.2500 54.7500 59.2500 63.7500 68.2500 72.7500 77.2500 81.7500
86.2500 90.7500

The vector xout shows that the center of the first bin is at 50.25, the center of the second bin is at 54.75 and so on.

5.8 POLAR PLOTS

Polar coordinates, in which the position of a point in a plane is defined by the angle θ and the radius (distance) to the point, are frequently used in the solution of science and engineering problems. The polar command is used to plot functions in polar coordinates. The command has the form:

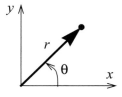

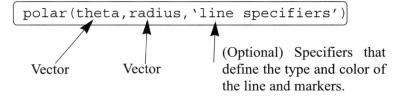

Vector   Vector

(Optional) Specifiers that define the type and color of the line and markers.

where theta and radius are vectors whose elements define the coordinates of the points to be plotted. The polar command plots the points and draws the polar grid. The line specifiers are the same as in the plot command. To plot a function $r = f(\theta)$ in a certain domain, a vector for values of θ is created first, and then a vector r with the corresponding values of $f(\theta)$ is created using element-by-element calculations. The two vectors are then used in the polar command.

For example, a plot of the function $r = 3\cos^2(0.5\theta) + \theta$ for $0 \le \theta \le 2\pi$ is shown below.

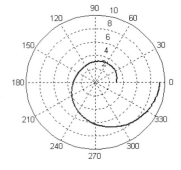

```
t = linspace(0,2*pi,200);
r = 3*cos(0.5*t).^2+t;
polar(t,r)
```

5.9 PLOTTING MULTIPLE PLOTS ON THE SAME PAGE

Multiple plots on the same page can be created with the subplot command, which has the form:

$$\boxed{\texttt{subplot(m,n,p)}}$$

The command divides the Figure Window (page when printed) into $m \times n$ rectangular subplots where plots will be created. The subplots are arranged like elements in a $m \times n$ matrix where each element is a subplot. The subplots are numbered from 1 through $m \cdot n$. The upper left is 1 and the lower right is the number $m \cdot n$. The numbers increase from left to right within a row, from the first row to the last. The command subplot(m,n,p) makes the subplot p current. This means that the next plot command (and any

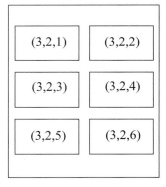

formatting commands) will create a plot (with the corresponding format) in this subplot. For example, the command subplot(3,2,1) creates 6 areas arranged in 3 rows and 2 columns as shown, and makes the upper left subplot current. An example of using the subplot command is shown in the solution of Sample Problem 5-2.

5.10 EXAMPLES OF MATLAB APPLICATIONS

Sample Problem 5-2:   Piston-crank mechanism

The piston-connecting rod-crank mechanism is used in many engineering applications. In the mechanism shown in the following figure, the crank is rotating at a constant speed of 500 rpm.

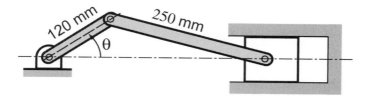

Calculate and plot the position, velocity, and acceleration of the piston for one revolution of the crank. Make the three plots on the same page. Set $\theta = 0°$ when $t = 0$.

Solution

The crank is rotating with a constant angular velocity $\dot{\theta}$. This means that if we set $\theta = 0°$ when $t = 0$, then at time t the angle θ is given by $\theta = \dot{\theta}t$, and that $\ddot{\theta} = 0$ at all times.

The distances d_1 and h are given by:

$$d_1 = r\cos\theta \quad \text{and} \quad h = r\sin\theta$$

Knowing h, the distance d_2 can be calculated using the Pythagorean theorem:

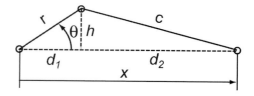

$$d_2 = (c^2 - h^2)^{1/2} = (c^2 - r^2\sin^2\theta)^{1/2}$$

The position x of the piston is then given by:

$$x = d_1 + d_2 = r\cos\theta + (c^2 - r^2\sin^2\theta)^{1/2}$$

The derivative of x with respect to time gives the velocity of the piston:

$$\dot{x} = -r\dot{\theta}\sin\theta - \frac{r^2\dot{\theta}\sin 2\theta}{2(c^2 - r^2\sin^2\theta)^{1/2}}$$

The second derivative of x with respect to time gives the acceleration of the piston:

$$\ddot{x} = -r\dot{\theta}^2\cos\theta - \frac{4r^2\dot{\theta}^2\cos 2\theta(c^2 - r^2\sin^2\theta) + (r^2\dot{\theta}\sin 2\theta)^2}{4(c^2 - r^2\sin^2\theta)^{3/2}}$$

In the equation above $\ddot{\theta}$ was taken to be zero.

A MATLAB program (script file) that calculates and plots the position, velocity, and acceleration of the piston for one revolution of the crank is shown below:

```
THDrpm = 500; r = 0.12; c = 0.25;                    Define θ, r and c.
THD = THDrpm*2*pi/60;              Change the units of θ̇ from rpm to rad/s.
tf = 2*pi/THD;                Calculate the time for one revolution of the crank.
t = linspace(0,tf,200);         Create a vector for the time with 200 elements.
TH = THD*t;                                       Calculate θ for each t.
d2s = c^2-r^2*sin(TH).^2;             Calculate d2 squared for each θ.
x = r*cos(TH) + sqrt(d2s);                 Calculate x for each θ.
xd = -r*THD*sin(TH) - (r^2*THD*sin(2*TH))./(2*sqrt(d2s));
xdd = -r*THD^2*cos(TH) - (4*r^2*THD^2*cos(2*TH).*d2s +
(r^2*sin(2*TH)*THD).^2)./(4*d2s.^(3/2));
                                       Calculate ẋ and ẍ for each θ.
subplot(3,1,1)
plot(t,x)                                              Plot x vs. t.
grid                                              Format the first plot.
xlabel('Time (s)')
ylabel('Position (m)')
subplot(3,1,2)
plot(t,xd)                                             Plot ẋ vs. t.
grid                                           Format the second plot.
xlabel('Time (s)')
ylabel('Velocity (m/s)')
subplot(3,1,3)
plot(t,xdd)                                            Plot ẍ vs. t.
grid                                              Format the third plot.
xlabel('Time (s)')
ylabel('Acceleration (m/s^2)')
```

When the script file runs it generates the three plots on the same page as shown in Figure 5-11. The figure nicely shows that the velocity of the piston is zero at the end points of the travel range where the piston changes the direction of the motion. The acceleration is maximum (directed to the left) when the piston is at the right end.

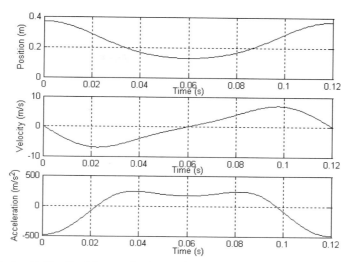

Figure 5-11: Position, velocity, and acceleration of the piston vs. time.

Sample Problem 5-3: Electric Dipole

The electric field at a point due to a charge is a vector **E** with magnitude E given by Coulomb's law:

$$E = \frac{1}{4\pi\varepsilon_0}\frac{q}{r^2}$$

where $\varepsilon_0 = 8.8541878 \times 10^{-12}\dfrac{C^2}{N \cdot m^2}$ is the permittivity constant, q is the magnitude of the charge, and r is the distance between the charge and the point. The direction of **E** is along the line that connects the charge with the point. **E** points outward from q if q is positive, and toward q if q is negative. An electric dipole is created when a positive charge and a negative charge of equal magnitude are placed some distance apart. The electric field, **E**, at any point is obtained by superposition of the electric field of each charge.

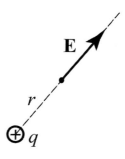

An electric dipole with $q = 12 \times 10^{-9}C$ is created, as shown in the figure. Determine and plot the magnitude of the electric field along the x-axis from $x = -5$cm to $x = 5$cm.

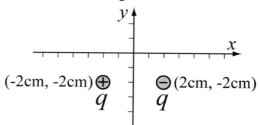

Solution

The electric field **E** at any point $(x, 0)$ along the x-axis is obtained by adding the electric field vectors due to each of the charges.

$$\mathbf{E} = \mathbf{E}_- + \mathbf{E}_+$$

The magnitude of the electric field is the length of the vector **E**.

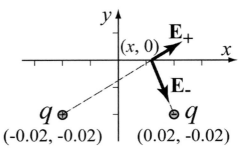

The problem is solved by following these steps:

Step 1: Create a vector x for points along the x-axis.

Step 2: Calculate the distance (and distance^2) from each charge to the points on the x-axis.

$$r_{minus} = \sqrt{(0.02 - x)^2 + 0.02^2} \qquad r_{plus} = \sqrt{(x + 0.02x)^2 + 0.02^2}$$

Step 3: Write unit vectors in the direction from each charge to the points on the x-axis.

$$\mathbf{E}_{minusUV} = \frac{1}{r_{minus}}((0.02 - x)\mathbf{i} - 0.02\mathbf{j})$$

$$\mathbf{E}_{plusUV} = \frac{1}{r_{plus}}((x + 0.02)\mathbf{i} + 0.02\mathbf{j})$$

Step 4: Calculate the magnitude of the vector $\mathbf{E}_-$ and $\mathbf{E}_+$ at each point by using Coulomb's law.

$$E_{minusMAG} = \frac{1}{4\pi\varepsilon_0}\frac{q}{r_{minus}^2} \qquad E_{plusMAG} = \frac{1}{4\pi\varepsilon_0}\frac{q}{r_{plus}^2}$$

Step 5: Create the vectors $\mathbf{E}_-$ and $\mathbf{E}_+$ by multiplying the unit vectors by the magnitudes.

Step 6: Create the vector **E** by adding the vectors $\mathbf{E}_-$ and $\mathbf{E}_+$.

Step 7: Calculate E, the magnitude (length) of **E**.

Step 8: Plot E as a function of x.

A program in a script file that solves the problem is:

```
q = 12e-9;
epsilon0 = 8.8541878e-12;
x = [-0.05:0.001:0.05]';
rminusS = (0.02-x).^2 + 0.02^2; rminus=sqrt(rminusS);
rplusS = (x + 0.02).^2 + 0.02^2; rplus=sqrt(rplusS);
```

Create a column vector x.

Step 2, each variable is a column vector.

```
EminusUV = [((0.02 - x)./rminus), (-0.02./rminus)];
EplusUV = [((x + 0.02)./rplus), (0.02./rplus)];
EminusMAG = (q/(4*pi*epsilon0))./rminusS;
EplusMAG = (q/(4*pi*epsilon0))./rplusS;
Eminus = [EminusMAG.*EminusUV(:,1), EminusMAG.*EminusUV(:,2)];
Eplus = [EplusMAG.*EplusUV(:,1), EplusMAG.*EplusUV(:,2)];
E = Eminus + Eplus;
EMAG = sqrt(E(:,1).^2 + E(:,2).^2);
plot(x,EMAG,'k','linewidth',1)
xlabel('Position along the x-axis (m)','FontSize',12)
ylabel('Magnitude of the electric field (N/C)','FontSize',12)
title('ELECTRIC FIELD DUE TO AN ELECTRIC DIPOLE','FontSize',12)
```

> Steps 3 & 4, each variable is a two column matrix. Each row is the vector for the corresponding x.

Step 6.

Step 7.

Step 5.

When this script file is executed in the Command Window the following figure is created in the Figure Window:

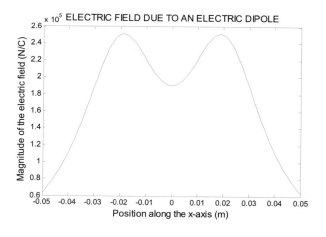

5.11 PROBLEMS

1. Make two separate plots of the function $f(x) = 0.6x^5 - 5x^3 + 9x + 2$; one plot for $-4 \le x \le 4$, and one for $-2.7 \le x \le 2.7$.

2. Plot the function $f(x) = \dfrac{x^2 - x + 1}{x^2 + x + 1}$ for $-10 \le x \le 10$.

3. Use the `fplot` command to plot the function:

$$f(x) = 0.01x^5 - 0.03x^4 + 0.4x^3 - 2x^2 - 6x + 5$$

in the domain $-4 \leq x \leq 6$.

4. Plot the function $f(x) = \dfrac{1.5x}{x-4}$ for $-10 \leq x \leq 10$. Notice that the function has a vertical asymptote at $x = 4$. Plot the function by creating two vectors for the domain of x. The first vector (call it $x1$) with elements from -10 to 3.7, and the second vector (call it $x2$) with elements from 4.3 to 10. For each of the x vector create a y vector (call them $y1$ and $y2$) with the corresponding values of y according to the function. To plot the function make two curves in the same plot ($y1$ vs. $x1$, and $y2$ vs. $x2$).

5. Plot the function $f(x) = \dfrac{x^2 - 5x + 10}{x^2 - 2x - 3}$ for $-10 \leq x \leq 10$. Notice that the function has two vertical asymptotes. Plot the function by dividing the domain of x into three parts; one from -10 to near the left asymptote, one between the two asymptotes, and one from near the right asymptote to 10. Set the range of the y-axis from -20 to 20.

6. Plot the function $f(x) = 3x\sin(x) - 2x$ and its derivative, both on the same plot, for $-2\pi \leq x \leq 2\pi$. Plot the function with a solid line, and the derivative with a dashed line. Add a legend and label the axes.

7. An electrical circuit that includes a voltage source v_S with an internal resistance r_S, and a load resistance R_L is shown in the figure. The power P dissipated in the load is given by:

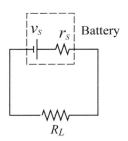

$$P = \dfrac{v_S^2 R_L}{(R_L + r_S)^2}$$

Plot the power P as a function of R_L for $1 \leq R_L \leq 10\ \Omega$, given that $v_S = 12$ V, and $r_S = 2.5\ \Omega$.

8. Ship A travels south at a speed of 8 miles/hour, and a ship B travels east at a speed of 16 miles/hour. At 7 AM the ships are positioned as shown in the figure. Plot the distance between the ships as a function of time for the next 4 hours. The horizontal axis should show the actual time of day starting at 7 AM, while the vertical axis shows the

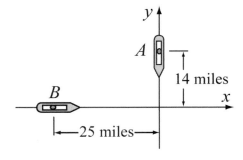

distance. Label the axes. If visibility is 8 miles, estimate from the graph the time when people from the two ships can see each other.

9. The value V of an initial investment P in a savings account paying an annual interest rate r is given by: $V = P\left(1 + \frac{r}{m}\right)^{mt}$, where m is the number of times the interest is compounded in a year, and t is the number of years. (If the interest is compounded annually $m = 1$, quarterly $m = 4$, and so on.) If the interest is compounded continuously, the value is given by: $V = Pe^{rt}$.

    Consider an investment of \$5,000 for 15 years at an annual interest rate of 7.5%. Show the difference in the value of the account when the interest is compounded annually, quarterly, and continuously by constructing a plot that shows the value of the investment as a function of time (years) for each compounding method. Plot the three cases in the same plot, use a different line types for each plot, label the axes, create a legend, and add a title to the plot.

10. The Gateway Arch in St. Louis is shaped according to the equation:

$$y = 693.8 - 68.8\cosh\left(\frac{x}{99.7}\right) \text{ ft.}$$

Make a plot of the Arch.

11. The magnitude M, on the Richter scale, of an earthquake is given by:

$$M = \frac{2}{3}\log\frac{E}{10^{4.4}}$$

where E is the energy in Joules released by the earthquake.
Make a plot of E (ordinate) versus M (abscissa) for $3 \le M \le 8$. Use a logarithmic scale for E and a linear scale for M. Label the axes and add a title to the plot.

12. The position x as a function of time of a particle that moves along a straight line is given by:

$$x(t) = 0.4t^3 - 2t^2 - 5t + 13 \text{ m}$$

The velocity $v(t)$ of the particle is determined by the derivative of $x(t)$ with respect to t, and the acceleration $a(t)$ is determined by the derivative of $v(t)$ with respect to t.

Derive the expressions for the velocity and acceleration of the particle, and make plots of the position, velocity, and acceleration as a function of time for $0 \le t \le 7$ s. Use the `subplot` command to make the three plots on the same page with the plot of the position on the top, the velocity at the middle, and the acceleration at the bottom. Label the axes appropriately with the correct units.

13. In a typical tension test a dog-bone-shaped specimen is pulled in a machine. During the test, the force F needed to pull the specimen and the length L of a gage section are

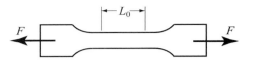

measured. This data is used for plotting a stress-strain diagram of the material. Two definitions, engineering and true, exist for stress and strain. The engineering stress σ_e and strain ε_e are defined by:

$$\sigma_e = \frac{F}{A_0} \text{ and } \varepsilon_e = \frac{L - L_0}{L_0}, \text{ where } L_0 \text{ and } A_0 \text{ are the initial gauge length and the}$$

initial cross-sectional area of the specimen, respectively. The true stress σ_t and strain ε_t are defined by:

$$\sigma_t = \frac{F L}{A_0 L_0} \text{ and } \varepsilon_t = \ln\frac{L}{L_0} .$$

    The following are measurements of force and gauge length from a tension test with an aluminum specimen. The specimen has a round cross section with radius of 6.4 mm (before the test). The initial gauge length is $L_0 = 25\text{mm}$. Use the data to calculate and plot the engineering and true stress-strain curves, both on the same plot of the material. Label the axes and label the curves. Units: When the force is measured in Newtons (N), and the area is calculated in m², the units of the stress are Pascals (Pa).

F (N)	0	13345	26689	40479	42703	43592	44482	44927
L (mm)	25	25.037	25.073	25.113	25.122	25.125	25.132	25.144
F (N)	45372	46276	47908	49035	50265	53213	56161	
L (mm)	25.164	25.208	25.409	25.646	26.084	27.398	29.150	

14. A resistor, $R = 4\ \Omega$, and an inductor, $L = 1.3$ H, are connected in a circuit to a voltage source as shown in Figure (a) (RL circuit). When the voltage source

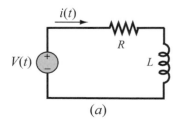

(a)

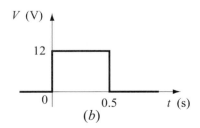

(b)

applies a rectangular voltage pulse with an amplitude of $V = 12$ V and a duration of 0.5 s, shown in Figure (b), The current $i(t)$ in the circuit as a function of time is given by:

$$i(t) = \frac{V}{R}(1 - e^{(-Rt)/L}) \quad \text{for } 0 \le t \le 0.5 \text{ s}$$

$$i(t) = e^{-(Rt)/L} \frac{V}{R}(e^{(0.5R)/L} - 1) \quad \text{for } 0.5 \le t \text{ s}$$

Make a plot of the current as a function of time for $0 \le t \le 2$ s.

15. The orbit of the planets around the sun can approximately be modeled by the polar equation:

$$r = \frac{eP}{1 - e\cos\theta}$$

The values of the constants P and e for four planets are given below. Plot the orbits of the four planets in one figure (use the `hold on` command).

Planet	P ($\times 10^6$ m)	e	Planet	P ($\times 10^6$ m)	e
Mercury	269.2	0.206	Earth	8964	0.0167
Venus	15913	0.00677	Mars	2421	0.0934

16. The radial probability density $P_r(r)$ for an excited hydrogen atom at the first excited state (quantum numbers $n = 2$ and $l = 0$) is given by:

$$P_r(r) = \frac{1}{8a_0}\left(\frac{r}{a_0}\right)^2\left(2 - \frac{r}{a_0}\right)e^{(-r/a_0)}$$

where $a_0 = 52.92 \times 10^{-11}$ m is the Bohr radius.
Make a plot of P_r as a function of r/a_0 for $0 \le r/a_0 \le 15$.

17. A cantilever beam is a beam that is clamped at one end and is free at the other end. The deflection y at point x of a beam loaded with a uniformly distributed load w is given by the equation:

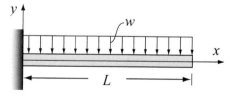

$$y = \frac{-w}{24EI}(x^4 - 4Lx^3 + 6L^2x^2)$$

where E is the elastic modulus, I is the moment of inertia, and L is the length of the beam. For the beam shown in the figure $L = 6$ m, $E = 70 \times 10^9$ Pa (aluminum), $I = 9.19 \times 10^{-6}$ m^4, and $w = 800$ N/m.
Make a plot of the deflection of the beam y as a function of x.

18. A simply supported beam that is subjected to a constant distributed load w over half of its length is shown in the figure. The deflection y, as a function of x, is given by the equations:

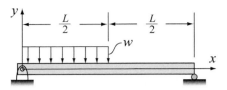

$$y = \frac{-wx}{384EI}(16x^3 - 24Lx^2 + 9L^3) \quad \text{for } 0 \le x \le \frac{L}{2}$$

$$y = \frac{-wL}{384EI}(8x^3 - 24Lx^2 + 17L^2x - L^3) \quad \text{for } \frac{L}{2} \le x \le L$$

where E is the elastic modulus, I is the moment of inertia, and L is the length of the beam. For the beam shown in the figure $L = 20$ m, $E = 200 \times 10^9$ Pa (steel), $I = 348 \times 10^{-6}$ m⁴, and $w = 5 \times 10^3$ N/m.
Make a plot of the deflection of the beam y as a function of x.

19. The yield stress (the limit of the elastic range) of most metals is sensitive to the rate at which the material is loaded. The data below gives the yield stress of a certain steel at various strain rates.

Strain Rate (s⁻¹)	0.00007	0.0002	0.05	0.8	4.2	215	3500
Yield Stress (MPa)	345	362	419	454	485	633	831

The yield stress as a function of strain rate can be modeled with the equation:

$$\sigma_{yld} = 350\left[\left(\frac{\varepsilon_t}{210}\right)^{0.16} + 1\right]$$

where σ_{yld} is the yield stress in MPa, and ε_t is the strain rate in s⁻¹. Make a plot of the yield stress (vertical axis) versus the strain rate (horizontal axis). Use linear scale for the stress and logarithmic scale for the strain rate. Show the data points as points with a marker and the model as a solid line. Label the axes and add a legend to the plot.

Chapter 6
Functions and Function Files

A simple function in mathematics, $f(x)$, associates a unique number to each value of x. The function can be expressed in the form $y = f(x)$, where $f(x)$ is usually a mathematical expression in terms of x. A value of y (output) is obtained when a value of x (input) is substituted in the expression. Many functions are programed inside MATLAB as built-in functions, and can be used in mathematical expressions simply by typing their name with an argument (see Section 1.5); examples are `sin(x)`, `cos(x)`, `sqrt(x)`, and `exp(x)`. Frequently, in computer programs, there is a need to calculate the value of functions that are not built-in. When a function expression is simple and needs to be calculated only once, it can be typed as part of the program. However, when a function needs to be evaluated many times for different values of arguments it is convenient to create a "user-defined" function. Once the new function is created (saved) it can be used just like the built-in functions.

A user-defined function is a MATLAB program that is created by the user, saved as a function file, and then can be used like a built-in function. The function can be a simple single mathematical expression or a complicated and involved series of calculations. In many cases it is actually a subprogram within a computer program. The main feature of a function file is that it has an input and an output. This means that the calculations inside the function file are carried out using the input data, and the results of the calculations are transferred out of the function file by the output. The input and the output can be one or several variables, and each can be a scalar, vector, or an array of any size. Schematically, a function file can be illustrated by:

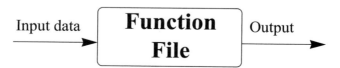

A very simple example of a user-defined function is a function that calculates the maximum height that a ball reaches when thrown up with a certain velocity. For a velocity v_0, the maximum height h_{max} is given by: $h_{max} = \dfrac{v_0^2}{2g}$, where g is the gravitational acceleration. In a function form it can be written as: $h_{max}(v_0) = \dfrac{v_0^2}{2g}$. In this case the input to the function is the velocity (a number), and the output is the maximum height (a number). For example, in SI units ($g = 9.81$ m/s^2) if the input is 15 m/s, the output is 11.47 m.

$$\xrightarrow{\text{15 m/s}} \boxed{\text{Function File}} \xrightarrow{\text{11.47 m}}$$

In addition to being used as a math function, function files can be used as subprograms in large programs. In this way large computer programs can be made up of smaller "building blocks" that can be tested independently. Function files are similar to subroutines in Basic and Fortran, procedures in Pascal, and functions in C.

Function files are explained in Sections 6.1 through 6.7. In addition to function files that are saved in separate files and are called for use in a computer program, MATLAB provides an option to define and use a user-defined math function within a computer program (not in a separate file). This can be done with the `inline` command which is explained in Section 6.8.

6.1 CREATING A FUNCTION FILE

Function files are created and edited, like script files, in the Editor/Debugger Window. This window is opened from the Command Window. In the **File** menu, select **New**, and then select **M-file**. Once the Editor/Debugger Window opens it looks like that shown in Figure 6-1, and the commands of the file can be typed line after line. The first line in a function file must be the function definition line as discussed in the next section.

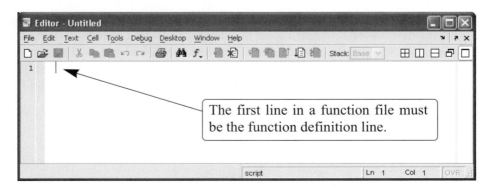

Figure 6-1: The Editor/Debugger Window.

6.2 STRUCTURE OF A FUNCTION FILE

The structure of a typical function file is shown in Figure 6-2. This particular function calculates the monthly payment and the total payment of a loan. The inputs to the function are the amount of the loan, the annual interest rate, and the duration of the loan (number of years of the loan). The output from the function is the monthly payment and the total payment.

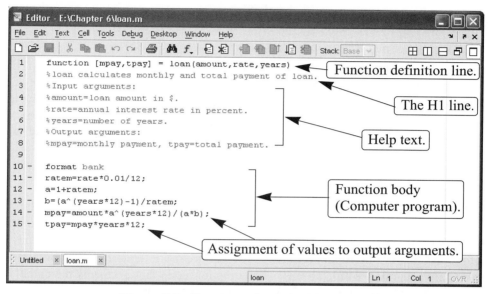

Figure 6-2: Structure of a typical function file.

The various parts of the function file are described in detail in the following sections.

6.2.1 Function Definition Line

The first executable line in a function file **must** be the function definition line. Otherwise the file is considered a script file. The function definition line:

• Defines the file as a function file.

• Defines the name of the function.

• Defines the number and order of the input and output arguments.

The form of the function definition line is:

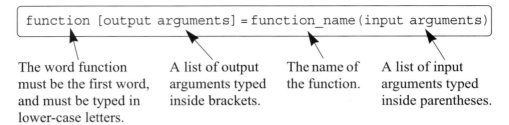

The word function
must be the first word,
and must be typed in
lower-case letters.

A list of output
arguments typed
inside brackets.

The name of
the function.

A list of input
arguments typed
inside parentheses.

The word function, typed in lower case letters, must be the first word in the function definition line. On the screen the word function appears in blue. The function name is typed following the equal sign. The name can be made up of letters, digits, and the underscore character. The rules for the name are the same as the rules for naming variables as described in Section 1.6.2. It is good practice to avoid names of built-in functions, and names of variables already defined by the user or predefined by MATLAB.

6.2.2 Input and Output Arguments

The input and output arguments are used to transfer data into and out of the function. The input arguments are listed inside parentheses following the function name. Usually, there is at least one input argument, although it is possible to have a function that has no input arguments. If there are more than one, the input arguments are separated with commas. The computer code that performs the calculations within the function file is written in terms of the input arguments and assumes that the arguments have assigned numerical values. This means that the mathematical expressions in the function file must be written according to the dimensions of the arguments, since the arguments can be scalars, vectors, and arrays. In the example shown in Figure 6-2 there are three input arguments, (amount, rate, years) and in the mathematical expressions they are assumed to be scalars. The actual values of the input arguments are assigned when the function is used (called). Similarly, if the input arguments are vectors or arrays, the mathematical expressions in the function body must be written to follow linear algebra or element-by-element calculations.

The output arguments, which are listed inside brackets on the left side of the assignment operator in the function definition line, transfer the output from the function file. Function files can have none, one, or several output arguments. If there are more than one, the output arguments are separated with commas. If there is only one output argument it can be typed without brackets. **In order for the function file to work, the output arguments must be assigned values in the computer program that is in the function body.** In the example in Figure 6-2 there are two output arguments, [mpay, tpay]. When a function does not have an output argument, the assignment operator in the function definition line can be omitted. A function without an output argument can, for example, generate a plot or print data to a file.

It is also possible to transfer strings into a function file. This is done by typing the string as part of the input variables (text enclosed in single quotes). Strings can be used to transfer names of other functions into the function file.

Usually, all the input to, and the output from, a function file are transferred through the input and output arguments. In addition, however, all the input and output features of script files are valid and can be used in function files. This means that any variable that is assigned a value in the code of the function file will be displayed on the screen unless a semicolon is typed at the end of the command. In addition, the input command can be used to input data interactively, and the disp, fprintf, and plot commands can be used to display information on the screen, save to a file, or plot figures just as in a script file. The following are examples of function definition lines with different combinations of input and output arguments.

Function definition line	**Comments**
function[mpay,tpay] = loan(amount,rate,years)	Three input arguments, two output arguments.
function [A] = RectArea(a,b)	Two input arguments, one output argument.
function A = RectArea(a,b)	Same as above, one output argument can be typed without the brackets.
function [V, S] = SphereVolArea(r)	One input variable, two output variables.
function trajectory(v,h,g)	Three input arguments, no output arguments.

6.2.3 The H1 Line and Help Text Lines

The H1 line and help text lines are comment lines (lines that begin with the percent % sign) following the function definition line. They are optional, but frequently used to provide information about the function. The H1 line is the first comment line and usually contains the name and a short definition of the function. When a user types (in the Command Window) lookfor a_word, MATLAB searches for a_word in the H1 lines of all the functions, and if a match is found, the H1 line that contains the match is displayed.

The help text lines are comment lines that follow the H1 line. These lines contain an explanation of the function and any instructions related to the input and output arguments. The comment lines that are typed between the function definition line and the first non-comment line (the H1 line and the help text) are displayed when the user types `help function_name` in the Command Window. This is true for MATLAB built-in functions as well as the user-defined functions. For example, for the function `loan` in Figure 6-2, if `help loan` is typed in the Command Window (make sure the current directory or the search path includes the directory where the file is saved), the display on the screen is:

>> help loan

loan calculates monthly and total payment of loan.

Input arguments:

amount = loan amount in $.

rate = annual interest rate in percent.

years = number of years.

Output arguments:

mpay = monthly payment, tpay = total payment.

A function file can include additional comment lines in the function body. These lines are ignored by the `help` command.

6.2.4 Function Body

The function body contains the computer program (code) that actually performs the computations. The code can use all MATLAB programming features. This includes calculations, assignments, any built-in or user-defined functions, flow control (conditional statements and loops) as explained in Chapter 7, comments, blank lines, and interactive input and output.

6.3 LOCAL AND GLOBAL VARIABLES

All the variables in a function file are local (the input and output arguments and any variables that are assigned values within the function file). This means that the variables are defined and recognized only inside the function file. When a function file is executed, MATLAB uses an area of memory that is separate from the workspace (the memory space of the Command Window and the script files). In a function file the input variables are assigned values each time the function is called. These variables are then used in the calculations within the function file. When the function file finishes its execution the value of the output arguments are transferred to the variables that were used when the function was called. All of this means that a function file can have variables with the same name as variables in the Command Window or in script files. The function file does not recognize variables with the same name that have been assigned values outside the function. The assignment of values to these variables in the function file will not change their assignment elsewhere.

Each function file has it own local variables which are not shared with other functions or with the workspace of the Command Window and the script files. It is possible, however, to make a variable common (recognized) in several different function files, and perhaps in the workspace too. This is done by declaring the variable global with the global command that has the form:

```
global variable_name
```

Several variables can be declared global by listing them, separated with spaces, in the global command. For example:

```
global GRAVITY_CONST FrictionCoefficient
```

- The variable has to be declared global in every function file that the user wants it to be recognized in. The variable is then common only to these files.

- The global command must appear before the variable is used. It is recommended to enter the global command at the top of the file.

- The global command has to be entered in the Command Window, or/and in a script file, for the variable to be recognized in the workspace.

- The variable value can be assigned, or reassigned, a value in any of the locations it is declared common.

- It is recommended to use long descriptive names (or use all capital letters) for global variables in order to distinguish between them and regular variables.

6.4 SAVING A FUNCTION FILE

A function file must be saved before it can be used. This is done, as with a script file, by choosing **Save as ...** from the **File** menu, selecting a location (many students save to an A drive or a zip drive), and entering the file name. It is highly recommended that the file is saved with a name that is identical to the function name in the function definition line. In this way the function is called (used) by using the function name. (If a function file is saved with a different name, the name it is saved under must be used when the function is called.) Function files are saved with the extension .m. Examples:

Function definition line	**File name**
function[mpay,tpay] = loan(amount,rate,years)	loan.m
function [A] = RectArea(a,b)	RectArea.m
function [V, S] = SphereVolArea(r)	SphereVolArea.m
function trajectory(v,h,g)	trajectory.m

6.5 USING A FUNCTION FILE

A user-defined function is used in the same way as a built-in function. The function can be called from the Command Window, from a script file, or from another function. To use the function file, the directory where it was saved must either be in the current directory or be in the search path (see Sections 4.3.1 and 4.3.2).

A function can be used by assigning its output to a variable (or variables), as a part of a mathematical expression, as an argument in another function, or just by typing its name in the Command Window or in a script file. In all cases the user must know exactly what the input and output arguments are. An input argument can be a number, a computable expression, or it can be a variable that has an assigned value. The arguments are assigned according to their position in the input and output argument lists in the function definition line.

Two of the ways that a function can be used are illustrated below with the user-defined loan function in Figure 6-2 that calculates the monthly and total payment (two output arguments) of a loan. The input arguments are the loan amount, annual interest rate, and the length (number of years) of the loan. In the first illustration the loan function is used with numbers as input arguments:

```
>> [month total] = loan(25000,7.5,4)
```
First argument is loan amount, second is interest rate, and third is number of years.

```
month =
    600.72
total =
    28834.47
```

In the second illustration the loan function is used with two preassigned variables and a number as the input arguments:

```
>> a = 70000;  b = 6.5;
```
Define variables a and b.

```
>> [x y] = loan(a,b,30)
```
Use a, b, and the number 30 for input arguments and x (monthly pay) and y (total pay) for output arguments.

```
x =
    440.06
y =
  158423.02
```

6.6 EXAMPLES OF SIMPLE FUNCTION FILES

Sample Problem 6-1:   Function of a math function

Write a function file (name it chp6one) for the function $f(x) = \dfrac{x^4\sqrt{3x+5}}{(x^2+1)^2}$. The
input to the function is x and the output is $f(x)$. Write the function such that x can
be a vector. Use the function to calculate:

a) $f(x)$ for $x = 6$.
b) $f(x)$ for $x = 1, 3, 5, 7, 9$, and 11.

Solution

The function file for the function $f(x)$ is:

```
function y = chp6one(x)                         Function definition line.
y = (x.^4.*sqrt(3*x + 5))./(x.^2 + 1).^2;       Assignment to output argument.
```

Note that the mathematical expression in the function file is written for element-
by-element calculations. In this way if x is a vector y will also be a vector. The
function is saved and then the search path is modified to include the directory
where the file was saved. As shown below the function is used in the Command
Window.

a)   Calculating the function for $x = 6$ can be done by typing chp6one(6) in
the Command Window, or by assigning the value of the function to a new variable:

```
>> chp6one(6)
ans =
   4.5401
>> F = chp6one(6)
F =
   4.5401
```

b)   To calculate the function for several values of x, a vector with the values of x
is first created, and then used for the argument of the function.

```
>> x = 1:2:11
x =
    1    3    5    7    9    11
>> chp6one(x)
ans =
   0.7071   3.0307   4.1347   4.8971   5.5197   6.0638
```

Another way is to type the vector x directly in the argument of the function.

```
>> H = chp6one([1:2:11])
H =
   0.7071   3.0307   4.1347   4.8971   5.5197   6.0638
```

Sample Problem 6-2: Converting temperature units

Write a user-defined function (name it FtoC) that converts temperature in degrees F to temperature in degrees C. Use the function to solve the following problem. The change in the length of an object, ΔL, due to a change in the temperature, ΔT, is given by: $\Delta L = \alpha L \Delta T$, where α is the coefficient of thermal expansion. Determine the change in the area of a rectangular (4.5 m by 2.25 m) aluminum ($\alpha = 23 \cdot 10^{-6}$ 1/°C) plate if the temperature changes from 40°F to 92°F.

Solution

A user-defined function that converts degrees F to degrees C is:

```
function C = FtoC(F)                          Function definition line.
%FtoC converts degrees F to degrees C
C = 5*(F - 32)./9;                            Assignment to output argument.
```

A script file (named Chapter6Example2) that calculates the change of the area of the plate due to the temperature is:

```
a1 = 4.5; b1 = 2.25; T1 = 40; T2 = 92; alpha = 23e-6;
deltaT = FtoC(T2) - FtoC(T1);          Using the FtoC function to calculate the
                                        temperature difference in degrees C.
a2 = a1 + alpha*a1*deltaT;             Calculating the new length.
b2 = b1 + alpha*b1*deltaT;             Calculating the new width.
AreaChange = a2*b2 - a1*b1;           Calculating the change in the area.
fprintf('The change in the area is %6.5f meters square.',AreaChange)
```

Executing the script file in the Command Window gives the solution:

```
>> Chapter6Example2
The change in the area is 0.01346 meters square.
```

6.7 COMPARISON BETWEEN SCRIPT FILES AND FUNCTION FILES

Students that are studying MATLAB for the first time sometimes have difficulty understanding exactly the differences between script and function files since, for many of the problems that they are asked to solve by using MATLAB, either types of files can be used. The similarities, and differences between script and function files are summarized below.

- Both script and function files are saved with the extension .m (that is why they are sometimes called M-files).

- The first line in a function file is the function definition line.

- The variables in a function file are local. The variables in a script file are recognized in the Command Window.

- Script files can use variables that have been defined in the workspace.

- Script files contain a sequence of MATLAB commands (statements).

- Function files can accept data through input arguments and can return data through output arguments.

- When a function file is saved, the name of the file should be the same as the name of the function.

6.8 INLINE FUNCTIONS

Function files can be used for simple mathematical functions, for large and complicated math functions that require extensive programming, and as subprograms in large computer programs. In cases when the value of a relatively simple mathematical function has to be determined many times within a program, MATLAB provides the option of using inline functions. An inline function is defined within the computer code (not as a separate file like a function file) and is then used in the code. Inline functions can be defined in any part of MATLAB.

Inline functions are created with the `inline` command according to the following format:

```
name = inline('math expression typed as a string')
```

- The mathematical expression can have one or several independent variables.

- Any letter except i and j can be used for the independent variables in the expression.

- The mathematical expression can include any built-in or user-defined functions.

- The expression must be written according to the dimension of the argument (element-by-element or linear algebra calculations).

- The expression can not include preassigned variables.

- Once the function is defined it can be used by typing its name and a value for the argument (or arguments) in parenthesis (see example below).

- The `inline` function can also be used as an argument in other functions.

Inline function with one independent variable:

For example, the function: $f(x) = \dfrac{e^{x^2}}{\sqrt{x^2 + 5}}$ can be defined (in the Command Window) as an inline function for x as a scalar by:

```
>> FA = inline('exp(x^2)/sqrt(x^2 + 5)')
FA =
    Inline function:
    FA(x) = exp(x^2)/sqrt(x^2 + 5)
>>
```

If a semicolon is not typed at the end, MATLAB responds by displaying the function. The function can then be used for different values of x, as shown below.

```
>> FA(2)
ans =
    18.1994
>> z = FA(3)

z =
    2.1656e+003
>>
```

If x is expected to be an array, and the function calculated for each element, then the function must be modified for element-by-element calculations.

```
>> FA = inline('exp(x.^2)./sqrt(x.^2 + 5)')
FA =
    Inline function:
    FA(x) = exp(x.^2)./sqrt(x.^2 + 5)
>> FA([1 0.5 2])                          Using a vector as the argument.
ans =
    1.1097   0.5604   18.1994
```

Inline function with several independent variables:

An inline function that has two or more independent variables can be written in the format shown or by using the following format:

```
name = inline('mathematical expression','arg1',
                               'arg2','arg3')
```

In the format shown here the order of the arguments in the function is defined. If the previous format is used, MATLAB arranges the arguments in alphabetical order. For example, the $f(x, y) = 2x^2 - 4xy + y^2$ can be defined as an inline function by:

```
>> HA = inline('2*x^2 - 4*x*y + y^2')
HA =
    Inline function:
    HA(x,y) = 2*x^2 - 4*x*y + y^2
```

Then, the function can be used for different values of x and y. For example, HA(2,3) gives:

```
>> HA(2,3)
ans =
   -7
```

Another example of using an inline function with several arguments is shown next in Sample Problem 6-3.

Sample Problem 6-3:   Distance between points in polar coordinates

Write an inline function that calculates the distance between two points in a plane when the position of the points is given in polar coordinates. Use the `inline` function to calculate the distance between point A (2, $\pi/6$) and point B (5, $3\pi/4$).

Solution

The distance between two points in polar coordinates can be calculated by using the law of cosines:

$$d = \sqrt{r_A^2 + r_B^2 - 2r_A r_B \cos(\theta_A - \theta_B)}$$

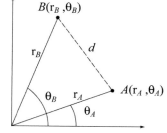

The formula for the distance is first entered as an inline function with four arguments. Then the function is used to calculate the distance between points A and B.

```
d = inline('sqrt(rA^2 + rB^2 - 2*rA*rB*cos(thetaB - thetaA))',...
'rA','thetaA','rB','thetaB')
```

> The order of the arguments is defined to be (rA, thetaA, rB, thetaB).

d =
   Inline function:
   d(rA,thetaA,rB,thetaB) = sqrt(rA^2 + rB^2 - 2*rA*rB*cos(thetaB - thetaA))
>> DistAtoB = d(2,pi/6,5,3*pi/4)
DistAtoB =
   5.8461 | The arguments are typed in the order defined in the function. |

6.9 THE feval COMMAND

The feval (function evaluate) command evaluates the value of a function for a given value (or values) of the function's argument (or arguments). The format of the command is:

| variable = feval('fuction name', argument value) |

The value that is determined by feval can be assigned to a variable, or if the command is typed without an assignment MATLAB displays ans = and the value of the function.

- The function name is typed as string.

- The function can be a built-in or a user-defined function.

- If there is more than one input argument, the arguments are separated with commas.

- If there is more than one output argument, the variables on the left-hand side of the assignment operator are typed inside brackets and separated with commas.

Two examples of using the feval command with built-in functions are:

>> feval('sqrt',64)

ans =
   8
>> x = feval('sin',pi/6)

x =
   0.5000

The following shows the use of the feval command with the user-defined function loan that was created earlier in the chapter (Figure 6-2):

>> [M,T] = feval('loan',50000,3.9,10) | A $50,000 loan, 3.9% interest, 10 years. |

M =
   502.22 | Monthly payment. |

T =
  60266.47

Total payment.

When to use `feval`:

In the examples above the outcome from using the `feval` command is the same as if the function was used directly. For the last example typing:

```
>> [M,T] = loan(50000,3.9,10)
M =
    502.22
T =
    60266.47
```

gives exactly the same result. It is obvious that there is no need to use the `feval` command to calculate a value of a function when the function itself can be used.

The `feval` command is useful in situations when the value of a function has to be calculated inside another function, and the inside function may need to be a different one when the outside function is called (used) at different times. This is done by importing the name (string) of the inside function through the input arguments of the outside function, and then using that string for the name of the function in the `feval` command.

For example, MATLAB has a built-in function, called `fzero`, that finds the zero of a function with one variable. `fzero` can be used to solve different functions. This is done by transferring the name of the function to be solved into the `fzero` function by the input arguments. `feval` is used in the computer code of `fzero` to determine values of the function that is solved. The `fzero` function is explained in more detail in Chapter 10.

6.10 EXAMPLES OF MATLAB APPLICATIONS

Sample Problem 6-4:   Exponential growth and decay

A model for exponential growth, or decay, of a quantity is given by:

$$A(t) = A_0 e^{kt}$$

where $A(t)$ and A_0 are the quantity at time t and time 0, respectively, and k is a constant unique to the specific application.

Write a user-defined function that uses this model to predict the quantity $A(t)$ at time t from knowing A_0 and $A(t_1)$ at some other time t_1. For function name and arguments use `At = expGD(A0,At1,t1,t)`, where the output argument `At` corresponds to $A(t)$, and the input arguments `A0,At1,t1,t` corresponds to A_0, $A(t_1)$, t_1, and t, respectively.

Use the function file in the Command Window for the following two cases:

a) The population of Mexico was 67 million in the year 1980 and 79 million in 1986. Estimate the population in 2000.

b) The half-life of a radioactive material is 5.8 years. How much of a 7-gram sample will be left after 30 years?

Solution

To use the exponential growth model, the value of the constant k has to be determined first by solving for k in terms of A_0, $A(t_1)$, and t_1:

$$k = \frac{1}{t_1}\ln\frac{A(t_1)}{A_0}$$

Once k is known, the model can be used to estimate the population at any time.
    The user-defined function that solves the problem is:

> function At = expGD(A0,At1,t1,t) | Function definition line. |
> % expGD calculates exponential growth and decay
> % Input arguments are:
> % A0: Quantity at time zero.
> % At1: Quantity at time t1.
> % t1: The time t1.
> % t: time t.
> % Output argument is:
> % At: Quantity at time t.
> k = log(At1/A0)/t1; | Determination of k. |
> At = A0*exp(k*t); | Determination of $A(t)$. (Assignment of value to output variable.) |

Once the function is saved, it is used in the Command Window to solve the two cases. For case *a)* $A_0 = 67$, $A(t_1) = 79$, $t_1 = 6$, and $t = 20$:

> \>> expGD(67,79,6,20)
> ans =
>   116.03 | Estimation of the population in the year 2000. |

For case *b)* $A_0 = 7$, $A(t_1) = 3.5$ (since t_1 corresponds to half-life, which is the time required for the material to decay to half of its initial quantity), $t_1 = 5.8$, and $t = 30$.

> \>> expGD(7,3.5,5.8,30)
> ans =
>   0.19 | The quantity of the material after 30 years. |

Sample Problem 6-5:   Motion of a Projectile

Create a function file that calculates the trajectory of a projectile. The inputs to the function are the initial velocity and the angle at which the projectile is fired. The outputs from the function are the maximum height and distance. In addition, the function generates a plot of the trajectory. Use the function to calculate the trajectory of a projectile that is fired at a velocity of 230 m/s at an angle of 39°.

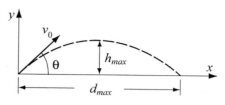

Solution

The motion of a projectile can be analyzed by considering the horizontal and vertical components. The initial velocity v_0 can be resolved into horizontal and vertical components:

$$v_{0x} = v_0\cos(\theta) \quad \text{and} \quad v_{0y} = v_0\sin(\theta)$$

In the vertical direction the velocity and position of the projectile are given by:

$$v_y = v_{0y} - gt \quad \text{and} \quad y = v_{0y}t - \frac{1}{2}gt^2$$

The time it takes the projectile to reach the highest point ($v_y = 0$) and the corresponding height are given by:

$$t_{hmax} = \frac{v_{0y}}{g} \quad \text{and} \quad h_{max} = \frac{v_{0y}^2}{2g}$$

The total flying time is twice the time it takes the projectile to reach the highest point, $t_{tot} = 2t_{hmax}$. In the horizontal direction the velocity is constant, and the position of the projectile is given by:

$$x = v_{0x}t$$

In MATLAB notation the function name and arguments are taken as: [hmax,dmax] = trajectory(v0,theta). The function file is:

function [hmax,dmax] = trajectory(v0,theta) | Function definition line. |

% trajectory calculates the max height and distance of a projectile, and makes a plot of the trajectory.

% Input arguments are:

% v0: initial velocity in (m/s).

% theta: angle in degrees.

% Output arguments are:

% hmax: maximum height in (m).

% dmax: maximum distance in (m).

```
% The function creates also a plot of the trajectory.
g = 9.81;
v0x=v0*cos(theta*pi/180);
v0y = v0*sin(theta*pi/180);
thmax = v0y/g;
hmax = v0y^2/(2*g);
ttot = 2*thmax;
dmax = v0x*ttot;
% Creating a trajectory plot
tplot = linspace(0,ttot,200);          Creating a time vector with 200 elements.
x = v0x*tplot;                          Calculating the x and y coordi-
y = v0y*tplot - 0.5*g*tplot.^2;         nates of the projectile at each time.
plot(x,y)                               Note the element-by-element multiplication.
xlabel('DISTANCE (m)')
ylabel('HEIGHT (m)')
title('PROJECTILE''S TRAJECTORY')
```

After the function is saved, it is used in the Command Window for a projectile that is fired at a velocity of 230 m/s, and an angle of 39°.

```
>> [h d] = trajectory(230,39)
h =
  1.0678e+003
d =
  5.2746e+003
```

In addition, the following figure is created in the Figure Window:

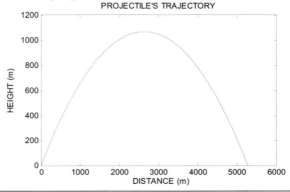

6.11 PROBLEMS

1. Write a user-defined MATLAB function, with two input and two output arguments that determines the height in centimeters and mass in kilograms of a person from his height in inches and weight in pounds. For the function name and arguments use [cm,kg] = STtoSI(in,lb). The input arguments are the height in inches and weight in pounds, and the output arguments are the height in centimeters and mass in kilograms. Use the function in the Command Window to:

   a) Determine in SI units the height and mass of a 5 ft. 10 in. person who weighs 175 lb.
   b) Determine your own height and weight in SI units.

2. Write a user-defined MATLAB function for the following math function:
$$y(x) = 0.9x^4 - 12x^2 - 5x$$
   The input to the function is x and the output is y. Write the function such that x can be a vector.
   a) Use the function to calculate $y(-3)$, and $y(5)$.
   b) Use the function to make a plot of the function $y(x)$ for $-4 \le x \le 4$.

3. Write a user-defined MATLAB function for the following function:
$$r(\theta) = 2(1.1 - \sin^2\theta)$$
   The input to the function is θ (in radians) and the output is r. Write the function such that θ can be a vector.
   a) Use the function to calculate $r(\pi/3)$, and $r(3\pi/2)$.
   b) Use the function to plot (polar plot) $r(\theta)$ for $0 \le \theta \le 2\pi$.

4. Write a user-defined MATLAB function that calculates the local maximum or minimum of a quadratic function of the form: $f(x) = ax^2 + bx + c$. For the function name and arguments use [x,y] = maxmin(a,b,c). The input arguments are the constants a, b, and c, and the output arguments are the coordinates x and y of the maximum or the minimum.
   Use the function to determine the maximum or minimum of the following functions:
   a) $f(x) = 3x^2 - 18x + 48$
   b) $f(x) = -5x^2 + 10x - 3$

5. The value P of a savings account with an initial investment of P_0, and annual interest rate r (in %), after t years is:
$$P = P_0\left(1 + \frac{r}{100}\right)^t$$

Write a user-defined MATLAB function that calculates the value of a savings account. For the function name and arguments use P = saval(PO,r,t). The inputs to the function are the initial investment, the interest rate, and the number of years. The output is the value of the account. Use the function to calculate the value of a $10,000 investment at an annual interest rate of 6% after 13 years.

6. Write a user-defined MATLAB function that converts torque given in units of lb-in. to torque in units of N-m. For the function name and arguments use Nm = lbintoNm(lbin). The input argument is the torque in lb-in., and the output argument is the torque in N-m. Use the function to convert 500 lb-in. to units of N-m.

7. Write a user-defined MATLAB function that determines the angles of a triangle when the lengths of the sides are given. For the function name and arguments use [al,bet,gam] = triangle(a,b,c). Use the function to determine the angles in triangles with the following sides:
   a) $a = 10, b = 15, c = 7$.
   b) $a = 6, b = 8, c = 10$.
   c) $a = 200, b = 75, c = 250$.

8. Write a user-defined MATLAB function that determines the unit vector in the direction of the line that connects two points (*A* and *B*) in space. For the function name and arguments use n = unitvec(A,B). The input to the function are two vectors A and B, each has three elements which are the Cartesian coordinates of the corresponding point. The output is a vector with the three components of the unit vector in the direction from *A* to *B*. Use the function to determine the following unit vectors:
   a) In the direction from point (2, 6, 5) to point (–10, 15, 9).
   b) In the direction from point (–10, 15, 9) to point (2, 6, 5).
   c) In the direction from point (1, 1, 2) to point (2, 1, 1).

9. The standard form of the equation of a straight line in the *x-y* plane is:
$Ax + By + C = 0$, and a point in the plane is defined by its coordinates (x_o, y_o).

Write a user-defined MATLAB function that determines the distance between a point and a straight line in the *x-y* plane. For the function name and arguments use:
d = PtoLdist(xo,yo,A,B,C)
where the input arguments are the coordinates of the point and the three constants of

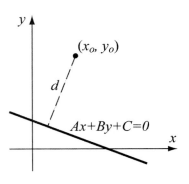

the equation of the line. The output argument is the distance. Use the function to determine the distance for the following cases:

a) Point: (2, –4), line: $-2x + 3.5y - 6 = 0$.

b) Point: (11, 2), line: $y = -2x + 6$, (note that the equation has the slope intercept form).

10. Write a user-defined MATLAB function that calculates a student's final grade in a course using the scores from two midterm exams, a final exam, and five homework assignments. The midterms are graded on a scale from 0 to 100, and are each 20% of the final grade. The final exam is graded on a scale from 0 to 100, and is 40% of the final grade. There are five homework assignments which are graded each on a scale from 0 to 10. The homework assignments together are 20% of the final grade.

For the function name and arguments use g = fgrade (R). The input argument R is a matrix in which the elements in each row are the grades of one student. The first five columns are the homework grades (numbers between 0 and 10), the next two columns are the midterm grades (numbers between 0 and 100), and the last column is the final exam grade (a number between 0 and 100). The output from the function, g, is a column vector with the final grades for the course. Each row has the final grade of the student with the grades in the corresponding row of the matrix R.

The function can be used to calculate the grades of any number of students. For one student the matrix R has one row. Use the function in the following cases:

a) Use the Command Window to calculate the grade of one student with the following grades: (10, 5, 8, 7, 9, 75, 87, 69).

b) Write a program in a script file. The program asks the user to enter the students' grades in an array (each student a row). The program then calculates the final grades by using the function fgrade. Run the script file in the Command Window to calculate the grades of the following four students:

Student A: 7, 9, 5, 8, 10, 90, 70, 85.
Student B: 6, 4, 7, 0, 7, 60, 71, 50.
Student C: 5, 9, 10, 3, 5, 45, 75, 80.
Student D: 8, 8, 7, 7, 9, 82, 81, 88.

11. When *n* electrical resistors are connected in parallel, their equivalent resistance R_{Eq} can be determined from:

$$\frac{1}{R_{Eq}} = \frac{1}{R_1} + \frac{1}{R_2} + \dots + \frac{1}{R_n}$$

Write a user-defined MATLAB function that calculates R_{Eq}. For the function name and arguments use REQ = req (R). The input to the function is a vector in which each element is a resistor value, and the output from the function is R_{Eq}. Use the function to calculate the equivalent resistance when the following resistors are connected in parallel: 50Ω, 75Ω, 300Ω, 60Ω, 500Ω, 180Ω, and 200Ω.

12. Write a user-defined MATLAB function that gives a random integer number within a range between two numbers. For the function name and arguments use n = randint (a,b), where the two input arguments a and b are the two numbers and the output argument n is the random number.

    Use the function in the Command Window for the following:

    a) Generate a random number between 1 and 49.

    b) Generate a random number between –35 and –2.

13. The area moment of inertia I_{x_o} of a rectangle about the axis x_o that passes through the centroid is: $I_{x_o} = \frac{1}{12}bh^3$. The moment of inertia about an axis x that is parallel to x_o is given by: $I_x = I_{x_o} + Ad_x^2$, where A is the area of the rectangle, and d_x is the distance between the two axes.

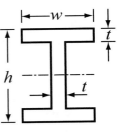

Write a MATLAB user-defined function that determines the area moment of inertia of an "I" beam about the axis that passes through its centroid (see drawing). For the function name and arguments use I = Ibeam(w,h,t). The inputs to the function are the width w, the height h, and the thickness t of the web and flanges.

(The moment of inertia of a composite area is obtained by dividing the area into parts and adding the moment of inertia of the parts.)

Use the function to determine the moment of inertia of an "I" beam with $w = 200$ mm, $h = 300$ mm, and $t = 22$ mm.

14. A two-dimensional state of stress at a point in a loaded material is defined by three components of stresses σ_{xx}, σ_{yy}, and τ_{xy}. The maximum and minimum normal stresses (principal stresses) at the point, σ_{max} and σ_{min}, are calculated from the stress components by:

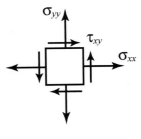

$$\sigma_{\substack{max \\ min}} = \frac{\sigma_{xx} + \sigma_{yy}}{2} \pm \sqrt{\left(\frac{\sigma_{xx} - \sigma_{yy}}{2}\right)^2 + \tau_{xy}^2}$$

Write a user-defined MATLAB function that determines the principal stresses from the stress components. For the function name and arguments use [Smax,Smin] = princstress(Sxx,Syy,Sxy). The input arguments are the three stress components, and the output arguments are the maximum and minimum stresses.

Use the function to determine the principal stresses for the following states of stress:

a) σ_{xx} = 150 MPa, σ_{yy} = –40 MPa, and τ_{xy} = 80 MPa.
b) σ_{xx} = –12 ksi, σ_{yy} = –16 ksi, and τ_{xy} = –7 ksi.

15. In a low-pass RC filter (a filter that passes signals with low frequencies), the ratio of the magnitude of the voltages is given by:

$$RV = \left|\frac{V_o}{V_i}\right| = \frac{1}{\sqrt{1+(\omega RC)^2}}$$

where ω is the frequency of the input signal.

Write a user-defined MATLAB function that calculates the magnitude ratio. For the function name and arguments use RV = lowpass(R,C,w). The input arguments are: R the size of the resistor in Ω (ohms), C the size of the capacitor in F (Farads), and w the frequency of the input signal in rad/s. Write the function such that w can be a vector.

Write a program in a script file that uses the lowpass function to generate a plot of RV as a function of ω for $10^{-2} \leq \omega \leq 10^{6}$ rad/s. The plot has a logarithmic scale on the horizontal axis (ω). When the script file is executed, it asks the user to enter the values of R and C. Label the axes of the plot.
Run the script file with R = 1200 Ω, and C = 8 μF.

16. A band-pass filter passes signals with frequencies within a certain range. In this filter the ratio of the magnitude of the voltages is given by:

$$RV = \left|\frac{V_o}{V_i}\right| = \frac{\omega RC}{\sqrt{(1-\omega^2 LC)^2+(\omega RC)^2}}$$

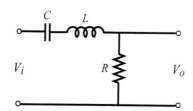

where ω is the frequency of the input signal.
Write a user-defined MATLAB function that calculates the magnitude ratio. For the function name and arguments use RV = bandpass(R,C,L,w). The input arguments are: R the size of the resistor in Ω (ohms), C the size of the capacitor in F (Farads), L is the inductance of the coil in H (Henrys), and w the frequency of the input signal in rad/s. Write the function such that w can be a vector.

Write a program in a script file that uses the bandpass function to generate a plot of RV as a function of ω for $10^{-2} \leq \omega \leq 10^{7}$ rad/s. The plot has a logarithmic scale on the horizontal axis (ω). When the script file is executed, it asks the user to enter the values of R, L, and C. Label the axes of the plot.

Run the script file for the following two cases:

a) $R = 1100\,\Omega$, and $C = 9\,\mu F$, $L = 7\,mH$.

b) $R = 500\,\Omega$, and $C = 300\mu F$, $L = 400\,mH$.

Chapter 7
Programming in MATLAB

A computer program is a sequence of computer commands. In a simple program the commands are executed one after the other in the order that they are typed. In this book, for example, all the programs that have been presented so far, in script or function files, are simple programs. Many situations, however, require more sophisticated programs in which commands are not necessarily executed in the order that they are typed, or that different commands (or groups of commands) are executed when the program runs with different input variables. For example, a computer program that calculates the cost of mailing a package for the post office uses different mathematical expressions to calculate the cost depending on the weight and size of the package, the content (books are less expensive to mail), and the type of service (airmail, ground, etc.). In other situations there might be a need to repeat a sequence of commands several times within a program. For example, programs that solve equations numerically repeat a sequence of calculations until the error in the answer is smaller than some measure.

MATLAB provides several tools that can be used to control the flow of a program. Conditional statements (Section 7.2) and the `switch` structure (Section 7.3), make it possible to skip commands or to execute specific groups of commands in different situations. `For` loops and `while` loops (Section 7.4) make it possible to repeat a sequence of commands several times.

It is obvious that changing the flow of a program requires some kind of decision-making process within the program. The computer must decide whether to execute the next command or to skip one or more commands and continue at a different line in the program. The program makes these decisions by comparing values of variables. This is done by using relational and logical operators which are explained in Section 7.1.

It should also be noted that function files (Chapter 6) can be used in programming. A function file is a subprogram. When the program reaches the command line that has the function, it provides input to the function, and "waits" for the results. The function carries out the calculations, transfers the results back to the program that "called" the function, which then continues to the next command.

7.1 RELATIONAL AND LOGICAL OPERATORS

A relational operator compares two numbers by determining whether a comparison statement (e.g. 5 < 8) is true or false. If the statement is true, it is assigned a value of 1. If the statement if false, it is assigned a value of 0. A logical operator examines true/false statements and produces a result which is true (1) or false (0) according to the specific operator. For example, the logical AND operator gives 1 only if both statements are true. Relational and logical operators can be used in mathematical expressions, and, as will be shown in this chapter, are frequently used in combination with other commands, to make decisions that control the flow of a computer program.

Relational Operators:

Relational operators in MATLAB are:

Relational Operator	**Description**
<	Less than.
>	Greater than.
<=	Less than or equal to.
>=	Greater than or equal to.
==	Equal to.
~=	Not Equal to.

Note that the "equal to" relational operator consists of two = signs (with no space between them), since one = sign is the assignment operator. Also, in other relational operators that consist of two characters there is no space between the characters (<=, >=, ~=)

- Relational operators are used as arithmetic operators within a mathematical expression. The result can be used in other mathematical operations, in addressing arrays, and together with other MATLAB commands (e.g. if) to control the flow of a program.

- When two numbers are compared, the result is 1 (logical true) if the comparison, according to the relational operator, is true, and 0 (logical false) if the comparison is false.

- If two scalars are compared, the result is a scalar 1 or 0. If two arrays are compared (only arrays with the same size can be compared), the comparison is done *element-by-element*, and the result is a logical array of the same size with 1's and 0's according to the outcome of the comparison at each address.

- If a scalar is compared with an array, the scalar is compared with every element of the array, and the result is a logical array with 1's and 0's according to the outcome of the comparison of each element.

Some examples are:

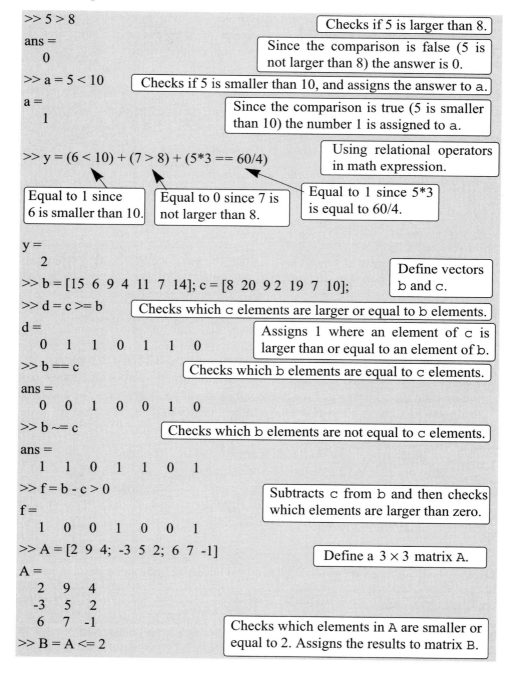

```
>> 5 > 8
```
Checks if 5 is larger than 8.

```
ans =
     0
```
Since the comparison is false (5 is not larger than 8) the answer is 0.

```
>> a = 5 < 10
```
Checks if 5 is smaller than 10, and assigns the answer to a.

```
a =
     1
```
Since the comparison is true (5 is smaller than 10) the number 1 is assigned to a.

```
>> y = (6 < 10) + (7 > 8) + (5*3 == 60/4)
```
Using relational operators in math expression.

Equal to 1 since 6 is smaller than 10.

Equal to 0 since 7 is not larger than 8.

Equal to 1 since 5*3 is equal to 60/4.

```
y =
     2
>> b = [15  6  9  4  11  7  14]; c = [8  20  9 2  19  7  10];
```
Define vectors b and c.

```
>> d = c >= b
```
Checks which c elements are larger or equal to b elements.

```
d =
     0   1   1   0   1   1   0
```
Assigns 1 where an element of c is larger than or equal to an element of b.

```
>> b == c
```
Checks which b elements are equal to c elements.

```
ans =
     0   0   1   0   0   1   0
>> b ~= c
```
Checks which b elements are not equal to c elements.

```
ans =
     1   1   0   1   1   0   1
>> f = b - c > 0
```
Subtracts c from b and then checks which elements are larger than zero.

```
f =
     1   0   0   1   0   0   1
>> A = [2 9 4; -3 5 2; 6 7 -1]
```
Define a 3 × 3 matrix A.

```
A =
     2   9   4
    -3   5   2
     6   7  -1
>> B = A <= 2
```
Checks which elements in A are smaller or equal to 2. Assigns the results to matrix B.

```
B =
    1   0   0
    1   0   1
    0   0   1
```

- The results of a relational operation with vectors, which are vectors with 0's and 1's, are called logical vectors and can be used for addressing vectors. When a logical vector is used for addressing another vector, it extracts from that vector the elements in the positions where the logical vector has 1's. For example:

```
>> r = [8 12 9 4 23 19 10]
```
Define a vector r.
```
r =
    8   12   9   4   23   19   10
>> s = r <= 10
```
Checks which r elements are smaller than or equal to 10.
```
s =
    1   0   1   1   0   0   1
```
A logical vector s with 1's at positions where elements of r are smaller than or equal to 10.
```
>> t = r(s)
```
Use s for addresses in vector r to create vector t.
```
t =
    8   9   4   10
```
Vector t consists of elements of r in positions where s has 1's.
```
>> w = r(r <= 10)
```
The same can be done in one step.
```
w =
    8   9   4   10
```

- Numerical vectors and arrays with the numbers 0's and 1's are not the same as logical vectors and arrays with 0's and 1's. Numerical vectors and arrays can not be used for addressing. Logical vectors and arrays, however, can be used in arithmetic operations. The first time a logical vector or an array is used in arithmetic operations it is changed to a numerical vector or array.

- Order of precedence: In a mathematical expression that includes relational and arithmetic operations, the arithmetic operations (+, −, *, /, \) have precedence over relational operations. The relational operators themselves have equal precedence and are evaluated from left to right. Parentheses can be used to alter the order of precedence. Examples are:

```
>> 3 + 4 < 16/2
```
+ and / are executed first.
```
ans =
    1
```
The answer is 1 since 7 < 8 is true.
```
>> 3 + (4 < 16)/2
```
4 < 16 is executed first, and is equal to 1, since it is true.
```
ans =
    3.5000
```
3.5 is obtained from 3 + 1/2.

Logical Operators:

Logical operators in MATLAB are:

Logical Operator	Name	Description
& Example: A&B	AND	Operates on two operands (A and B). If both are true, the result is true (1), otherwise the result is false (0).
| Example: A|B	OR	Operates on two operands (A and B). If either one, or both are true, the result is true (1), otherwise (both are false) the result is false (0).
~ Example: ~A	NOT	Operates on one operand (A). Gives the opposite of the operand. True (1) if the operand is false, and false (0) if the operand is true.

- Logical operators have numbers as operands. A nonzero number is true, and a zero number is false.

- Logical operators (like relational operators) are used as arithmetic operators within a mathematical expression. The result can be used in other mathematical operations, in addressing arrays, and together with other MATLAB commands (e.g. `if`) to control the flow of a program.

- Logical operators (like relational operators) can be used with scalars and arrays.

- The logical operations AND and OR can have both operands as scalars, arrays, or one array and one a scalar. If both are scalars, the result is a scalar 0 or 1. If both are arrays they must be of the same size and the logical operation is done *element-by-element*. The result is an array of the same size with 1's and 0's according to the outcome of the operation at each position. If one operand is a scalar and the other is an array, the logical operation is done between the scalar and each of the elements in the array and the outcome is an array of the same size with 1's and 0's.

- The logical operation NOT has one operand. When it is used with a scalar the outcome is a scalar 0 or 1. When it is used with an array, the outcome is an array of the same size with 1's in positions where the array has nonzero numbers and 0's in positions where the array has zeros.

The following are some examples:

```
>> 3&7
```
3 AND 7.

```
ans =
    1
>> a = 5|0
a =
    1
>> ~25
ans =
    0
>> t = 25*((12&0) + (~0) + (0|5))
t =
   50
>> x = [9 3 0 11 0 15]; y = [2 0 13 -11 0 4];
>> x&y
ans =
    1    0    0    1    0    1
>> z = x|y
z =
    1    1    1    1    0    1
>> ~(x + y)
ans =
    0    0    0    1    1    0
```

> 3 and 7 are both true (nonzero), so the outcome is 1.

> 5 OR 0 (assign to variable a).

> 1 is assigned to a since at least one number is true (nonzero).

> NOT 25.

> The outcome is 0 since 25 is true (nonzero) and the opposite is false.

> Using logical operators in math expression.

> Define two vectors x and y.

> The outcome is a vector with 1 in every position that both x and y are true (nonzero elements), and 0's otherwise.

> The outcome is a vector with 1 in every position that either or both x and y are true (nonzero elements), and 0's otherwise.

> The outcome is a vector with 0 in every position that the vector x + y is true (nonzero elements), and 1 in every position that x + y is false (zero element).

Order of precedence:

Arithmetic, relational, and logical operators can all be combined in mathematical expressions. When an expression has such a combination, the result depends on the order in which the operations are carried out. The following is the order used by MATLAB:

Precedence	Operation
1 (highest)	Parentheses (If nested parentheses exist, inner have precedence).
2	Exponentiation.
3	Logical NOT (~).
4	Multiplication, division.
5	Addition, subtraction.
6	Relational operators (>, <, >=, <=, ==, ~=).

Precedence	Operation
7	Logical AND (&).
8 (lowest)	Logical OR (|).

If two or more operations have the same precedence, the expression is executed in order from left to right.

It should be pointed out here that the order shown above is used since MATLAB 6. Previous versions of MATLAB have a slightly different order (& did not have precedence over |), and the user must be careful. Compatibility problems between different versions of MATLAB can be avoided by using parentheses even when they are not required.

The following are examples of expressions that include arithmetic, relational, and logical operators:

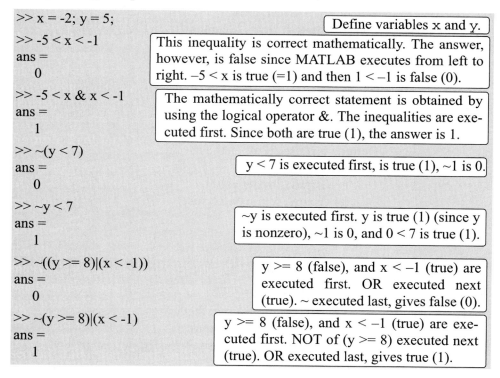

Built-in Logical Functions:

MATLAB has built-in functions that are equivalent to the logical operators. These functions are:

and(A,B)	equivalent to A&B
or(A,B)	equivalent to A|B
not(A)	equivalent to ~A

In addition, MATLAB has other logical built-in functions, some of which are described in the following table:

Function	Description	Example
`xor(a,b)`	Exclusive or. Returns true (1) if one operand is true and the other is false.	>> xor(7,0) ans = 1 >> xor(7,-5) ans = 0
`all(A)`	Returns 1 (true) if all elements in a vector A are true (nonzero). Returns 0 (false) if one or more elements are false (zero). If A is a matrix, treats columns of A as vectors, returns a vector with 1's and 0's.	>> A = [6 2 15 9 7 11]; >> all(A) ans = 1 >> B = [6 2 15 9 0 11]; >> all(B) ans = 0
`any(A)`	Returns 1 (true) if any element in a vector A is true (nonzero). Returns 0 (false) if all elements are false (zero). If A is a matrix, treats columns of A as vectors, returns a vector with 1's and 0's.	>> A = [6 0 15 0 0 11]; >> any(A) ans = 1 >> B = [0 0 0 0 0 0]; >> any(B) ans = 0

`find(A)`
 	
`find(A>d)`
	If A is a vector, returns the indices of the nonzero elements.
	
	If A is a vector, returns the address of the elements that are larger than d (any relational operator can be used).
		>> A = [0 9 4 3 7 0 0 1 8];
		>> find(A)
		ans =
		  2 3 4 5 8 9
		>> find(A > 4)
		ans =
		  2 5 9

The operation of the four logical operators, and, or, xor, and not can be summarized in a truth table:

INPUT		OUTPUT				
A	B	AND A&B	OR A|B	XOR (A,B)	NOT ~A	NOT ~B
false	false	false	false	false	true	true
false	true	false	true	true	true	false
true	false	false	true	true	false	true
true	true	true	true	false	false	false

Sample Problem 7-1:   Analysis of temperature data

The following were the daily maximum temperatures (in °F) in Washington DC during the month of April, 2002: 58 73 73 53 50 48 56 73 73 66 69 63 74 82 84 91 93 89 91 80 59 69 56 64 63 66 64 74 63 69, (data from the U.S. National Oceanic and Atmospheric Administration). Use relational and logical operations to determine the following:

a)   The number of days the temperature was above 75°.

b)   The number of days the temperature was between 65° and 80°.

c)   The days of the month that the temperature was between 50° and 60°.

Solution

In the script file below the temperatures are entered in a vector. Relational and logical expressions are then used to analyze the data.

```
T = [58  73  73  53  50  48  56  73  73  66  69  63  74  82  84 ...
     91  93  89  91  80  59  69  56  64  63 66  64  74  63  69];
Tabove75 = T >= 75;
```
A vector with 1's at addresses where T >= 75.

```
NdaysTabove75 = sum(Tabove75)
```
Add all the 1's in the vector `Tabove75`.

```
Tbetween65and80 = (T >= 65)&(T <= 80);
```
A vector with 1's at addresses where T >= 65 and T <= 80.

```
NdaysTbetween65and80 = sum(Tbetween65and80)
```
Add all the 1's in the vector `Tbetween65and80`.

```
datesTbetween50and60 = find((T >= 50)&(T <= 60))
```
The function `find` returns the address of the elements in T that have values between 50 and 60.

The script file (saved as Exp7_1) is executed in the Command Window:

```
>> Exp7_1
NdaysTabove75 =                              For 7 days the temp was above 75.
   7
NdaysTbetween65and80 =              For 12 days the temp was between 65 and 80.
  12
datesTbetween50and60 =                         Dates of the month with
    1    4    5    7   21   23                 temp between 50 and 60.
```

7.2 CONDITIONAL STATEMENTS

A conditional statement is a command that allows MATLAB to make a decision of whether to execute a group of commands that follow the conditional statement, or to skip these commands. In a conditional statement a conditional expression is stated. If the expression is true, a group of commands that follow the statement is executed. If the expression is false, the computer skips the group. The basic form of a conditional statement is:

> if conditional expression consisting of relational and/or logical operators.

Examples:

```
        if  a < b
        if  c >= 5
        if  a == b            All the variables must
        if  a ~= 0            have assigned values.
        if  (d<h) & (x>7)
        if  (x~=13) | (y<0)
```

- Conditional statements can be a part of a program written in a script file or a function file.

- As shown below, for every if statement there is an end statement.

     The if statement is commonly used in three structures, if-end, if-else-end, and if-elseif-else-end, which are described next.

7.2.1 The if-end Structure

The if-end structure of the conditional statement is shown schematically in Figure 7-1. The figure shows how the commands are typed in the program, and a flowchart that symbolically shows the flow, or the sequence, in which the commands are executed. As the program executes, it reaches the if statement. If the

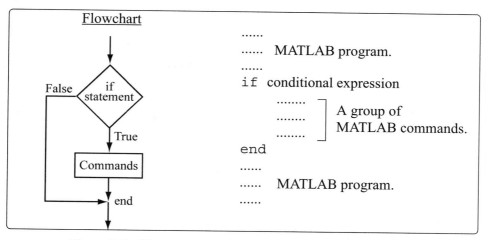

Figure 7-1: The structure of the `if-end` **conditional statement.**

conditional expression in the `if` statement is true (1), the program continues to execute the commands that follow the `if` statement all the way down to the `end` statement. If the conditional expression is false (0), the program skips the group of commands between the `if` and the `end`, and continues with the commands that follow the `end`.

The words `if` and `end` appear on the screen in blue, and the commands between the `if` statement and the `end` statement are automatically indented (they don't have to be indented), which makes the program easier to read. An example where the `if-end` statement is used in a script file is shown in Example 7-2.

Sample Problem 7-2:   Calculating worker's pay

A worker is paid according to his hourly wage up to 40 hours, and 50% more for overtime. Write a program in a script file that calculates the pay to a worker. The program asks the user to enter the number of hours and the hourly wage. The program then displays the pay.

Solution

The program in a script file is shown below. The program first calculates the pay by multiplying the number of hours by the hourly wage. Then an `if` statement checks whether the number of hours is greater than 40. If yes, the next line is executed and the extra pay for the hours above 40 is added. If not, the program skips to the `end`.

```
t = input('Please enter the number of hours worked  ');
h = input('Please enter the hourly wage in $  ');
Pay = t*h;
```

```
if t > 40
   Pay = Pay + (t - 40)*0.5*h;
end
fprintf('The worker''s pay is  $ %5.2f',Pay)
```

Application of the program (in the Command Window) for two cases is shown below (the file was saved as Workerpay):

```
>> Workerpay
Please enter the number of hours worked  35
Please enter the hourly wage in $  8
The worker's pay is  $ 280.00
>> Workerpay
Please enter the number of hours worked  50
Please enter the hourly wage in $  10
The worker's pay is  $ 550.00
```

7.2.2 The if-else-end *Structure*

This structure provides a means for choosing one group of commands, out of a possible two groups, for execution. The if-else-end structure is shown in Figure 7-2. The figure shows how the commands are typed in the program, and a

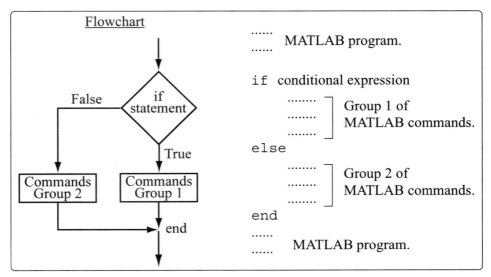

Figure 7-2: The structure of the if-else-end conditional statement.

flowchart that illustrates the flow, or the sequence, in which the commands are executed. The first line is an `if` statement with a conditional expression. If the conditional expression is true, the program executes group 1 of commands between the `if` and the `else` statements and then skips to the `end`. If the conditional expression is false, the program skips to the `else`, and then executes group 2 of commands between the `else` and the `end`.

The following example uses the `if-else-end` structure in a function file.

Sample Problem 7-3: Water level in water tower

The tank in a water tower has the geometry shown in the figure (the lower part is a cylinder and the upper part is an inverted frustum cone). Inside the tank there is a float that indicates the level of the water. Write a user-defined function file that determines the volume of the water in the tank from the position (height h) of the float. The input to the function is the value of h in m, and the output is the volume of the water in m^3.

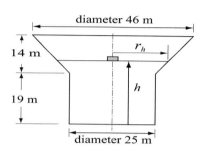

Solution

For $0 \le h \le 19$ m the volume of the water is given by the volume of a cylinder with height h: $V = \pi 12.5^2 h$.

For $19 < h \le 33$ m the volume of the water is given by adding the volume of a cylinder with $h = 19$ m, and the volume of the water in the cone:

$$V = \pi 12.5^2 \cdot 19 + \frac{1}{3}\pi(h-19)(12.5^2 + 12.5 \cdot r_h + r_h^2)$$

where r_h is: $r_h = 12.5 + \frac{10.5}{14}(h-19)$

The function, shown below, is named $v = $ `watervol(h)`.

```
function v = watervol(h)
% watervol calculates the volume of the water in the water tower.
% The input is the water level in meters.
% The output is the water volume in cubic meters.
if h <= 19
    v = pi*12.5^2*h;
else
    rh = 12.5 + 10.5*(h - 19)/14;
```

```
   v = pi*12.5^2*19 + pi*(h - 19)*(12.5^2 + 12.5*rh + rh^2)/3;
end
```

Two examples of using the function (in the Command Window) are given next:

```
>> watervol(8)
ans =
   3.9270e+003
>> VOL = watervol(25.7)
VOL =
   1.4115e+004
```

7.2.3 *The* `if-elseif-else-end` *Structure*

The `if-elseif-else-end` structure is shown in Figure 7-3. The figure shows how the commands are typed in the program, and a flowchart that illustrates the flow, or the sequence, in which the commands are executed. This structure includes two conditional statements (`if` and `elseif`) which make it possible to select one out of three groups of commands for execution. The first

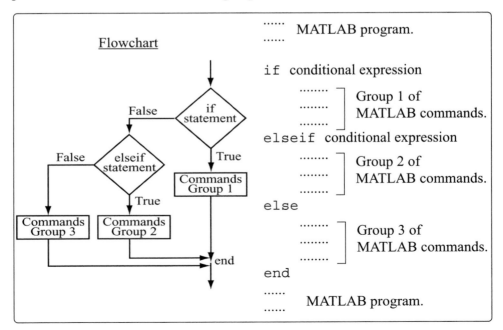

Figure 7-3: The structure of the `if-elseif-else-end` **conditional statement.**

line is an `if` statement with a conditional expression. If the conditional expression is true, the program executes group 1 of commands between the `if` and the `elseif` statement and then skips to the `end`. If the conditional expression in the `if` statement is false, the program skips to the `elseif` statement. If the conditional expression in the `elseif` statement is true the program executes group 2 of commands between the `elseif` and the `else` and then skips to the `end`. If the conditional expression in the `elseif` statement is false the program skips to the `else` and executes group 3 of commands between the `else` and the `end`.

It should be pointed out here that several `elseif` statements and associated groups of commands can be added. In this way more conditions can be included. Also, the `else` statement is optional. This means that in the case of several `elseif` statements and no `else` statement, if any of the conditional statements is true the associated commands are executed, but otherwise nothing is executed.

7.3 *THE* `switch-case` *STATEMENT*

The `switch-case` statement is another method that can be used to affect the flow of a program. It provides a means for choosing one group of commands for execution out of several possible groups. The structure of the statement is shown in Figure 7-4.

```
......          MATLAB program.
......

switch  switch expression
     case  value1
     ........  ] Group 1 of commands.
     ........
     case  value2
     ........  ] Group 2 of commands.
     ........
     case  value3
     ........  ] Group 3 of commands.
     ........
     otherwise
     ........  ] Group 4 of commands.
     ........
end
......          MATLAB program.
......
```

Figure 7-4: The structure of a `switch-case` **statement.**

- The first line is the switch command which has the form:

| switch   switch expression |

The switch expression can be a scalar or a string. Usually it is a variable that has an assigned scalar or a string. It can also be, however, a mathematical expression that includes preassigned variables and can be evaluated.

- Following the switch command there are one or several case commands. Each has a value (can be a scalar or a string) next to it (value1, value2, etc.) and associated group of commands below it.

- After the last case command there is an optional otherwise command followed by a group of commands.

- The last line must be an end statement.

How does the switch-case statement work?

The value of the switch expression in the switch command is compared with the values that are next to each of the case statements. If a match is found, the group of commands that follow the case statement with the match are executed. (Only one group of commands, the one between the case that matches and either the case, otherwise or end statement that is next, is executed).

- If there is more than one match, only the first matching case is executed.

- If no match is found and the otherwise statement (which is optional) exists, the group of commands between otherwise and end is executed.

- If no match is found and the otherwise statement does not exist, none of the command groups is executed.

- A case statement can have more than one value. This is done by typing the values in the form: {value1, value2, value3, ...}. (This form, which is not covered in this book, is called a cell array). The case is executed if at least one of the values matches the value of switch expression.

A NOTE: In MATLAB only the first matching case is executed. After the group of commands that are associated with the first matching are executed, the program skips to the end statement. This is different than the C language where break statements are required.

Sample Problem 7-4:   Converting units of energy

Write a program in a script file that converts a quantity of energy (work) given in units of either Joule, ft-lb, cal, or eV to the equivalent quantity in different units specified by the user. The program asks the user to enter the quantity of energy, its current units, and the new desired units. The output is the quantity of energy in the new units.

The conversion factors are: $1\,J\ =\ 0.738\ \text{ft-lb}\ =\ 0.239\ \text{cal}\ =\ 6.24{\times}10^{18}\ \text{eV}$.
Use the program to:

a) Convert 150 J to ft-lb.

b) 2800 cal to Joules.

c) 2.7 eV to cal.

Solution

The program includes two sets of `switch-case` statements and one `if-else-end` statement. The first `switch-case` statement is used to convert the input quantity from its initial units to units of Joules. The second is used to convert the quantity from Joules to the specified new units. The `if-else-end` statement is used to generate an error message if units are entered incorrectly.

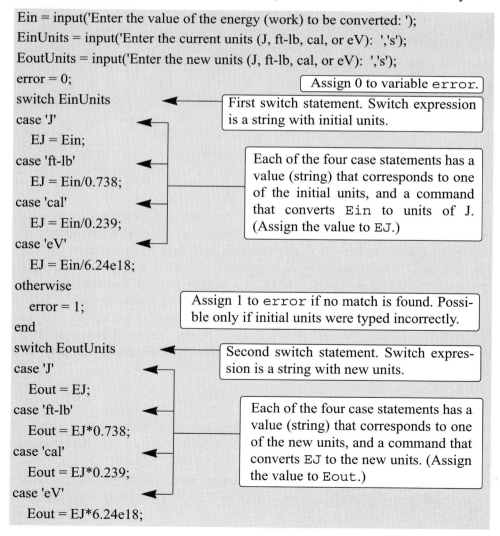

```
Ein = input('Enter the value of the energy (work) to be converted: ');
EinUnits = input('Enter the current units (J, ft-lb, cal, or eV): ','s');
EoutUnits = input('Enter the new units (J, ft-lb, cal, or eV): ','s');
error = 0;
switch EinUnits
case 'J'
    EJ = Ein;
case 'ft-lb'
    EJ = Ein/0.738;
case 'cal'
    EJ = Ein/0.239;
case 'eV'
    EJ = Ein/6.24e18;
otherwise
    error = 1;
end
switch EoutUnits
case 'J'
    Eout = EJ;
case 'ft-lb'
    Eout = EJ*0.738;
case 'cal'
    Eout = EJ*0.239;
case 'eV'
    Eout = EJ*6.24e18;
```

Assign 0 to variable `error`.

First switch statement. Switch expression is a string with initial units.

Each of the four case statements has a value (string) that corresponds to one of the initial units, and a command that converts `Ein` to units of J. (Assign the value to `EJ`.)

Assign 1 to `error` if no match is found. Possible only if initial units were typed incorrectly.

Second switch statement. Switch expression is a string with new units.

Each of the four case statements has a value (string) that corresponds to one of the new units, and a command that converts `EJ` to the new units. (Assign the value to `Eout`.)

otherwise
   error = 1;

> Assign 1 to `error` if no match is found. Possible only if new units were typed incorrectly.

end
if error   `If-else-if statement.`

> If `error` is true (nonzero), display an error message.

   disp('ERROR current or new units are typed incorrectly.')
else
   fprintf('E = %g %s',Eout,EoutUnits)   ◄─────────┐

> If `error` is false (zero), display converted energy.

end

As an example, the script file (saved as EnergyConversion) is used next in the Command Window to make the conversion in part *b* of the problem statement.

>> EnergyConversion
Enter the value of the energy (work) to be converted: 2800
Enter the current units (J, ft-lb, cal, or eV): cal
Enter the new units (J, ft-lb, cal, or eV): J
E = 11715.5 J

7.4 LOOPS

A loop is another method to alter the flow of a computer program. In a loop, the execution of a command, or a group of commands, is repeated several times consecutively. Each round of execution is called a pass. In each pass at least one variable, but usually more than one, or even all the variables that are defined within the loop are assigned new values. MATLAB has two kinds of loops. In `for-end` loops (Section 7.4.1) the number of passes is specified when the loop starts. In `while-end` loops (Section 7.4.2) the number of passes is not known ahead of time, and the looping process continues until a specified condition is satisfied. Both kinds of loops can be terminated at any time with the break command (see Section 7.6).

7.4.1 `for-end` *Loops*

In `for-end` loops the execution of a command, or a group of commands, is repeated a predetermined number of times. The form of the loop is shown in Figure 7-5.

- The loop index variable can have any variable name (usually `i`, `j`, `k`, `m`, and `n` are used. `i` and `j` should not be used if MATLAB is used with complex numbers).

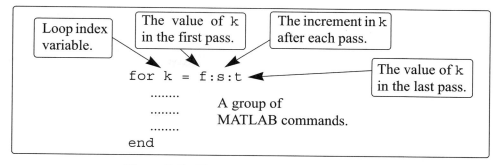

Figure 7-5: The structure of a `for-end` **loop.**

- In the first pass k = f and the computer executes the commands between the `for` and the `end` commands. Then, the program goes back to the `for` command for the second pass. k obtains a new value equal to k = f + s, and the commands between the `for` and the `end` commands are executed with the new value of k. The process repeats itself until the last pass where k = t. Then the program does not go back to the `for`, but continues with the commands that follow the `end` command. For example, if k = 1:2:9, there are five loops, and the value of k in each loop is 1, 3, 5, 7, and 9.

- The increment s can be negative (i.e. k = 25:–5:10 produces four passes with k = 25, 20, 15, 10).

- If the increment value s is omitted, the value is 1 (default) (i.e. k = 3:7 produces five passes with k = 3, 4, 5, 6, 7).

- If f = t, the loop is executed once.

- If f > t and s > 0, or if f < t and s < 0 the loop is not executed.

- If the values of k, s, and t are such that k cannot be equal to t, then, if s is positive, the last pass is the one where k has the largest value that is smaller than t (i.e. k = 8:10:50 produces five passes with k = 8, 18, 28, 38, 48). If s is negative the last pass is the one where k has the smallest value that is larger than t.

- In the `for` command k can also be assigned specific value (typed in as a vector). For example: for k = [7 9 –1 3 3 5].

- The value of k should not be redefined within the loop.

- Each `for` command in a program **must** have an `end` command.

- The value of the loop index variable (k) is not displayed automatically. It is possible to display the value in each pass (sometimes useful for debugging) by typing k as one of the commands in the loop.

- When the loop ends, the loop index variable (k) has the value that was last assigned to it.

A simple example of a for-end loop (in a script file) is:

```
for k = 1:3:10
    x = k^2
end
```

When this program is executed, the loop is executed four times. The value of k in the four passes is k = 1, 4, 7, and 10, which means that the values that are assigned to x in each of the passes are x = 1, 16, 49, and 100, respectively. Since a semicolon is not typed at the end of the second line, the value of x is displayed in the Command Window at each pass. When the script file is executed, the display in the Command Window is:

```
>> x =
    1
x =
    16
x =
    49
x =
    100
```

Sample Problem 7-5:   Sum of series

a) Use a for-end loop in a script file to calculate the sum of the first n terms of the series: $\sum_{k=1}^{n} \frac{(-1)^k k}{2^k}$. Execute the script file for $n = 4$ and $n = 20$.

b) The function sin(x) can be written as a Taylor series by:

$$\sin x = \sum_{k=0}^{\infty} \frac{(-1)^k x^{2k+1}}{(2k+1)!}$$

Write a user-defined function file that calculates sin(x) by using the Taylor's series. For the function name and arguments use y = Tsin(x,n). The input arguments are the angle x in degrees, and n the number of terms in the series. Use the function to calculate sin($150°$) using 3 and 7 terms.

Solution

a)   A script file that calculates the sum of the first n terms of the series is shown below.

```
n = input('Enter the number of terms ' );
S = 0;
for k = 1:n
    S = S + (-1)^k*k/2^k;
end
fprintf('The sum of the series is: %f',S)
```

Setting the sum to zero.

for-end loop.

In each pass one element of the series is calculated and is added to the sum of the elements from the previous passes.

The summation is done with a loop. In each pass one term of the series is calculated (in the first pass the first term, in the second pass the second term, and so on), and is added to the sum of the previous elements. The file is saved as Exp7-4a and then executed twice in the Command Window:

```
>> Exp7_4a
Enter the number of terms 4
The sum of the series is: -0.125000
>> Exp7_4a
Enter the number of terms 20
The sum of the series is: -0.222216
```

b)   A user-defined function file that calculates sin(*x*) by adding *n* terms of Taylor's formula is shown below.

```
function y = Tsin(x,n)
% Tsin calculates the sin using Taylor formula.
% Input arguments:
% x The angle in degrees, n number of terms.
xr = x*pi/180;
y = 0;
for k = 0:n - 1
    y = y + (-1)^k*xr^(2*k + 1)/factorial(2*k + 1);
end
```

Converting the angle from degrees to radians.

for-end loop.

The first element corresponds to $k = 0$ which means that in order to add *n* terms of the series, in the last loop $k = n - 1$. The function is used in the Command Window to calculate the sin(150°) using 3 and 7 terms:

```
>> Tsin(150,3)
```

Calculating sin(150°) with 3 terms of Taylor series.

```
ans =
    0.6523
>> Tsin(150,7)
ans =
    0.5000
```

Calculating sin(150°) with 7 terms of Taylor series.

The exact value is 0.5.

A note about `for-end` loops and element-by-element operations:

In some situations the same end result can be obtained by either using `for-end` loops or by using element-by-element operations. Sample Problem 7-5 illustrates how the `for-end` loop works, but the problem can also be solved by using element-by-element operations (see Problems 7 and 8 in Section 3.9). Element-by-element operations with arrays are one of the superior features of MATLAB that provides the means for computing in circumstances that otherwise require loops. In general, element-by-element operations are faster than loops and are recommended when either method can be used.

Sample Problem 7-6:   Modify vector elements

A vector is given by: $V = [5, 17, -3, 8, 0, -1, 12, 15, 20, -6, 6, 4, -7, 16]$. Write a program as a script file that doubles the elements that are positive and are divisible by 3 and/or 5, and raise to the power of 3 the elements that are negative but greater than –5.

Solution

The problem is solved by using a `for-end` loop that has an `if-elseif-end` conditional statement inside. The number of passes is equal to the number of elements in the vector. In each pass one element is checked by the conditional statement. The element is changed if it satisfies the conditions in the problem statement. A program in a script file that carries out the required operations is:

```
V = [5, 17, -3, 8, 0, -7, 12, 15, 20 -6, 6, 4, -2, 16];
n = length(V);
for k = 1:n
    if V(k) > 0 & (rem(V(k),3) == 0 | rem(V(k),5) == 0)
        V(k) = 2*V(k);
    elseif V(k) < 0 & V(k) > -5
        V(k) = V(k)^3;
    end
end
V
```

Setting n to be equal to the number of elements in V.

for-end loop.

if-elseif-end statement.

The program is executed in the Command Window:

```
>> Exp7_5
V =
   10   17   -27   8   0   -7   24   30   40   -6   12   4   -8   16
```

7.4.2 `while-end` *Loops*

`while-end` loops are used in situations when looping is needed but the number of passes is not known ahead of time. In `while-end` loops the number of passes is not specified when the looping process starts. Instead, the looping process continues until a stated condition is satisfied. The structure of a `while-end` loop is shown in Figure 7-6.

```
while  conditional expression
   ........
   ........              A group of
   ........              MATLAB commands.
end
```

Figure 7-6: The structure of a `while-end` loop.

The first line is a `while` statement that includes a conditional expression. When the program reaches this line the conditional expression is checked. If it is false (0), MATLAB skips to the `end` statement and continues with the program. If the conditional expression is true (1), MATLAB executes the group of commands that follow between the `while` and `end` command. Then MATLAB jumps back to the `while` command and checks the conditional expression. This looping process continues until the conditional expression is false.

For a `while-end` loop to execute properly:

- The conditional expression in the `while` command must include at least one variable.

- The variables in the conditional expression must have assigned values when MATLAB executes the `while` command for the first time.

- At least one of the variables in the conditional expression must be assigned a new value in the commands that are between the `while` and the `end`. Otherwise, once the looping starts it will never stop since the conditional expression will remain true.

An example of a simple `while-end` loop is shown in the following program. In this program a variable x with an initial value of 1 is doubled in each pass as long as its value is equal to or smaller than 15.

```
x = 1                        Initial value of x is 1.
while x <= 15                The next command is executed only if x <= 15.
    x = 2*x                  In each pass x doubles.
end
```

When this program is executed the display in the Command Window is:

```
x =
    1                        Initial value of x.
x =
    2
x =
    4                        In each pass x doubles.
x =
    8
x =
    16                       When x = 16, the conditional expression in the
                             while command is false and the looping stops.
```

Important note:

When writing a while-end loop, the programmer has to be sure that the variable (or variables) that are in the conditional expression and are assigned new values during the looping process will eventually be assigned values that make the conditional expression in the while command false. Otherwise the looping will continue indefinitely (indefinite loop). For example, in the example above if the conditional expression is changed to be x >= 0.5, the looping will continue indefinitely. Such a situation can be avoided by counting the passes and stopping the looping if the number of passes exceeds some large value. This can be done by adding the maximum number of passes to the conditional expression, or by using the break command (Section 7.6).

Since no one is free from making mistakes, a situation of indefinite loop can occur in spite of careful programming. If this happens, the user can stop the execution of an indefinite loop by pressing the **Ctrl + C** or **Ctrl + Break** keys.

Sample Problem 7-7:   Taylor series representation of a function

The function $f(x) = e^x$ can be represented in a Taylor series by $e^x = \sum_{n=0}^{\infty} \frac{x^n}{n!}$.

Write a program in a script file that determines e^x by using the Taylor series representation. The program calculates e^x by adding terms of the series and stopping when the absolute value of the term that was added last is smaller than 0.0001.

Use a `while-end` loop, but limit the number of passes to 30. If in the 30th pass the value of the term that is added is not smaller than 0.0001, the program stops and displays a message that more than 30 terms are needed.

Use the program to calculate e^2, e^{-4}, and e^{21}.

Solution

The first few terms of the Taylor series are:

$$e^x = 1 + x + \frac{x^2}{2!} + \frac{x^3}{3!} + \dots$$

A program that uses the series to calculate the function is shown next. The program asks the user to enter the value of x. Then, the first term an, is assigned the number 1, and an is assigned to the sum S. Then, from the second term and on, the program uses a `while` loop to calculate the nth term of the series and add it to the sum. The program also counts the number of terms n. The conditional expression in the `while` command is true as long as the absolute value of the nth term an is larger than 0.0001, and the number of passes n is smaller than 30. This means that if the 30th term is not smaller than 0.0001, the looping stops.

```
x = input('Enter x ');
n = 1; an = 1; S = an;
while abs(an) >= 0.0001 & n <= 30          Start of the while loop.
    an = x^n/factorial(n);                 Calculating the nth term.
    S = S + an;                            Adding the nth term to the sum.
    n = n + 1;                             Counting the number of passes.
end                                        end of the while loop.
if n >= 30                                 if-else-end loop.
    disp('More than 30 terms are needed')
else
fprintf('exp(%f) = %f',x,S)
fprintf('\nThe number of terms used is: %i',n)
end
```

The program uses an `if-else-end` statement to display the results. If the looping stopped because the 30th term is not smaller than 0.0001, it displays a message indicating this. If the value of the function is calculated successfully, it displays the value of the function and the number of terms used. When the program executes, the number of passes depends on the value of `x`. The program (saved as `expox`) is used to calculate e^2, e^{-4}, and e^{21}:

```
>> expox
Enter x 2                                    Calculating exp(2).
exp(2.000000) = 7.389046
The number of terms used is: 12              12 terms used.
>> expox
Enter x -4                                   Calculating exp(-4).
exp(-4.000000) = 0.018307
The number of terms used is: 18              18 terms used.
>> expox
Enter x 21                                   Trying to calculate exp(21).
More than 30 terms are needed
```

7.5 *Nested Loops and Nested Conditional Statements*

Loops and conditional statements can be nested within themselves and each other. This means that a loop and/or a conditional statement can start (and end) within another loop and/or conditional statement. There is no limit to the number of loops and conditional statements that can be nested. It must be remembered, however, that each `if`, `case`, `for`, and `while` statement must have a corresponding `end` statement. Figure 7-7 shows the structure of a nested `for-end` loop within

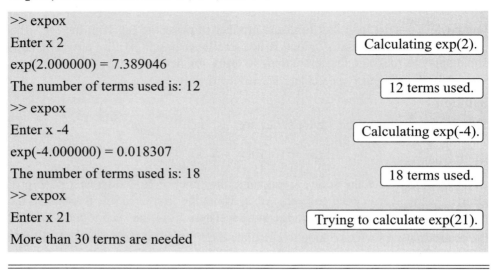

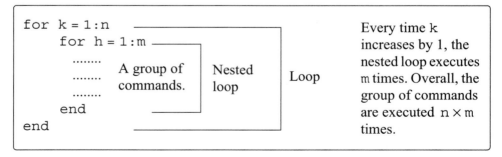

Figure 7-7: Structure of nested loops.

another `for-end` loop. In the loops shown in this figure, if, for example, `n` = 3 and `m` = 4, then first `k` = 1 and the nested loop executes four times with `h` = 1, 2, 3, and 4. Next `k` = 2 and the nested loop executes again four times with `h` = 1, 2, 3, and 4. Lastly `k` = 3 and the nested loop executes again four times. Every time a nested loop is typed MATLAB automatically indents the new loop relative to the outside loop. Nested loops and conditional statements are demonstrated in the following sample problem.

Sample Problem 7-8:   Creating a matrix with a loop

Write a program in a script file that creates a $n \times m$ matrix with elements that have the following values. The value of the elements in the first row is the number of the column. The value of the element in the first column is the number of the row. The rest of the elements are equal to the sum of the element above them and the element to the left. When executed, the program asks the user to enter values for n and m.

Solution

The program, shown below, has two loops (one nested), and a nested `if-elseif-else-end` statement. The elements in the matrix are assigned values row by row. The loop index variable of the first loop, `k`, is the address of the row, and the loop index variable of the second loop, `h`, is the address of the column.

```
n = input('Enter the number of rows ');
m = input('Enter the number of columns ');
A = [];                          Define an empty matrix A
for k = 1:n                      Start of the first for-end loop.
   for h = 1:m                   Start of the second for-end loop.
      if k == 1                  Start of the conditional statement.
         A(k,h) = h;             Assign values to the elements of the first row.
      elseif h == 1
         A(k,h) = k;             Assign values to the elements of the first column.
      else
         A(k,h) = A(k,h - 1) + A(k - 1,h);   Assign values to other elements.
      end                        end of the if statement.
   end                           end of the nested for-end loop.
end                              end of the first for-end loop.
A
```

The program is executed in the Command Window to create a 4×5 matrix.

```
>> Chap7_exp7
Enter the number of rows 4
```

Enter the number of columns 5
A =

1	2	3	4	5
2	4	7	11	16
3	7	14	25	41
4	11	25	50	91

7.6 *THE* break *AND* continue *COMMANDS*

The break **command:**

- When inside a loop (for and while), the break command terminates the execution of the loop (the whole loop, not just the last pass). When the break command appears in a loop, MATLAB jumps to the end command of the loop and continues with the next command (does not go back to the for command of that loop).

- If the break command is inside a nested loop, only the nested loop is terminated.

- When a break command appears outside a loop in a script, or function file, it terminates the execution of the file.

- The break command is usually used within a conditional statement. In loops it provides a method to terminate the looping process if some condition is met. For example, if the number of loops exceeds a predetermined value, or an error in some numerical procedure is smaller than a predetermined value. When typed outside a loop, the break command provides a means to terminate the execution of a file, such as if data transferred into a function file is not consistent with what is expected.

The continue **command:**

- The continue command can be used inside a loop (for and while) to stop the present pass and start the next pass in the looping process.

- The continue command is usually a part of a conditional statement. When MATLAB reaches the continue command, it does not execute the remaining commands in the loop, but skips to the end command of the loop and then starts a new pass.

7.7 EXAMPLES OF MATLAB APPLICATIONS

Sample Problem 7-9: Withdrawing from a retirement account.

A person in retirement is depositing $300,000 in a saving account which pays 5% interest per year. The person plans to withdraw money from the account once a year. He starts by withdrawing $25,000 after the first year, and then in future years, he increases the amount he withdraws according to the inflation rate. For example, if the inflation rate is 3%, he withdraws $25,750 after the second year. Calculate the number of years that the money in the account will last assuming a constant yearly inflation rate of 2%. Make a plot that shows the yearly withdrawal and the balance of the account over the years.

Solution

The problem is solved by using a loop (a `while` loop since the number of passes in not known before the loop starts). In each pass the amount to be withdrawn and account balance are calculated. The looping continues as long as the account balance is larger than or equal to the amount to be withdrawn. The following is a program in a script file that solves the problem. In the program, `year` is a vector in which each element is a year number, `W` is a vector with the amount withdrawn each year, and `AB` is a vector with the account balance each year.

```
rate = 0.05; inf = 0.02;
clear W AB year
year(1) = 0;                                    First element is year 0.
W(1) = 0;                                       Initial withdrawal amount.
AB(1) = 300000;                                 Initial account balance.
Wnext = 25000;                          The amount to be withdrawn after a year.
ABnext = 300000*(1 + rate);               The account balance after a year.
n = 2;
    while ABnext >= Wnext                 while checks if the next balance
        year(n) = n - 1;                 is larger then the next withdrawal.
        W(n) = Wnext;                       Amount withdrawn in year n – 1.
        AB(n) = ABnext - W(n);     Account balance in year n – 1 after withdrawal.
        ABnext = AB(n)*(1 + rate);   The account balance after additional year.
        Wnext = W(n)*(1 + inf);           The amount to be withdrawn
        n = n + 1;                           after additional year.
    end
fprintf('The money will last for %f years',year(n-1))
```

bar(year,[AB' W'],2.0)

The program is executed in the Command Window:

>> Chap7_exp9
The money will last for 15 years.

The program also generates the following figure (axes labels and legend were added to the plot by using the plot editor).

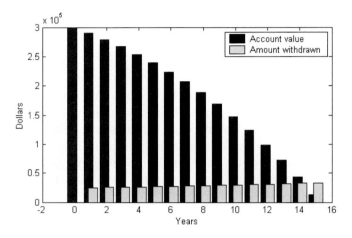

Sample Problem 7-10: Choosing numbers for a lottery

In a lottery the player has to select several numbers out of a list. Write a user-defined function that generates a list of n integer numbers that are uniformly distributed between the numbers a and b. All the selected numbers on the list must be different.

a) Use the function to generate a list of 6 numbers from the numbers 1 through 49.

b) Use the function to generate a list of 8 numbers from the numbers 60 through 75.

c) Use the function to generate a list of 9 numbers from the numbers −15 through 15.

Solution

The function, shown below, uses the MATLAB rand function (refer back to Section 3.7). To make sure all the numbers are different from each other, the numbers are picked one by one. Each number that is picked (by the rand function) is compared with all the numbers that were picked before. If a match is found, the number is not selected, and a new number is picked.

The while loop checks that the number does not match any of the elements (pre-

```
function x = lotto(a,b,n)
% lotto selects n numbers (all different) from the domain a,b.
% x is a vector with the n numbers.
for p = 1:n
    number = round((b - a)*rand + a);
    if p == 1
        x(p) = number;
    else
        r = 0;
        while r == 0
            r = 1;
            for k = 1:p - 1
                if x(k) == number
                    number = round((b - a)*rand + a);
                    r = 0;
                    break
                end
            end
        end
    end
    x(p) = number;
end
```

`number = round((b - a)*rand + a);`	Select an integer between *a* and *b*.
`x(p) = number;`	If the number is the first, it is assigned to $x(1)$.
`else`	If the number is not the first, it is compared to previous numbers.
`r = 0;`	Set r to zero.
`while r == 0`	See explanation below.
`r = 1;`	Set r to 1.
`for k = 1:p - 1`	`for` loop compares the new number to the ones in x (p).
`number = round((b - a)*rand + a); r = 0;`	If a match if found, a new number is selected and r is set to zero.
`break`	The last `for` loop is stopped. The program goes back to the `while` loop. Since r = 0 the loop inside the `while` starts again with the new number.
`x(p) = number;`	The number that does not match any previous numbers is assigned to x (p).

vious numbers) in the vector x. If a match is found, it keeps choosing new numbers until the new number is different than all the elements in x.

The function is used next in the Command Window for the three cases stated in the problem statement.

```
>> lotto(1,49,6)
ans =
    23    2   40   22   31   39
>> lotto(60,75,8)
ans =
    67   66   73   68   63   70   60   72
>> lotto(-15,15,9)
```

ans =

 -13   -4   0   -2   2   3   -12   12   8

Sample Problem 7-11: Flight of a model rocket

The flight of a model rocket can be modeled as follows. During the first 0.15 s the rocket is propelled up by the rocket engine with a force of 16 N. The rocket then flies up while slowing down under the force of gravity. After it reaches the apex, the rocket starts to fall back down. When its down velocity reaches 20 m/s a parachute opens (assumed to open instantly) and the rocket continues to move down at a constant speed of 20 m/s until it hits the ground. Write a program that calculates and plots the speed and altitude of the rocket as a function of time during the flight.

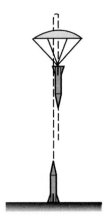

Solution

The rocket is assumed to be a particle that moves along a straight line in the vertical plane. For motion with constant acceleration along a straight line, the velocity and position as a function of time are given by:

$$v(t) = v_0 + at \quad \text{and} \quad s(t) = s_0 + v_0 t + \frac{1}{2}at$$

where v_0 and s_0 are the initial velocity and position, respectively. In the computer program the flight of the rocket is divided into three segments. Each segment is calculated in a while loop. In every pass the time increases by an increment.

Segment 1: The first 0.15 s when the rocket engine is on. During this period, the rocket moves up with a constant acceleration. The acceleration is determined by drawing a free body and a mass acceleration diagram (shown on the right). From Newton's second law, the sum of the forces in the vertical direction is equal to the mass times the acceleration (equilibrium equation):

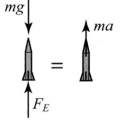

$$+\uparrow \Sigma F = F_E - mg = ma$$

Solving the equation for the acceleration gives:

$$a = \frac{F_E - mg}{m}$$

The velocity and height as a function of time are:

$$v(t) = 0 + at \quad \text{and} \quad h(t) = 0 + 0 + \frac{1}{2}at^2$$

where the initial velocity and initial position are both zero. In the computer program this segment starts when $t = 0$, and the looping continues as long as $t < 0.15$ s. The time, velocity, and height at the end of this segment are t_1, v_1, and h_1.

Segment 2: The motion from when the engine stops until the parachute opens. In this segment the rocket moves with a constant deceleration g. The speed and height of the rocket as a function of time are given by:

$$v(t) = v_1 - g(t - t_1) \quad \text{and} \quad h(t) = h_1 + v_1(t - t_1) - \frac{1}{2}g(t - t_1)^2$$

In this segment the looping continues until the velocity of the rocket is -20 m/s (negative since the rocket moves down). The time and height at the end of this segment are t_2, and h_2.

Segment 3: The motion from when the parachute opens until the rocket hits the ground. In this segment the rocket moves with constant velocity (zero acceleration). The height as a function of time is given by: $h(t) = h_2 - v_{chute}(t - t_2)$, where v_{chute} is the constant velocity after the parachute opens. In this segment the looping continues as long as the height is greater than zero.

A program in a script file that carries out the calculations is shown below:

```
m = 0.05; g = 9.81; tEngine = 0.15; Force = 16; vChute = -20; Dt = 0.01;
clear t v h
n = 1;
t(n) = 0; v(n) = 0; h(n) = 0;
% Segment 1
a1 = (Force - m*g)/m;
while t(n) < tEngine & n < 50000          The first while loop.
   n = n + 1;
   t(n) = t(n - 1) + Dt;
   v(n) = a1*t(n);
   h(n) = 0.5*a1*t(n)^2;
end
v1 = v(n); h1 = h(n); t1 = t(n);
% Segment 2
while v(n) >= vChute & n < 50000          The second while loop.
   n = n + 1;
   t(n) = t(n - 1) + Dt;
   v(n) = v1 - g*(t(n) - t1);
   h(n) = h1 + v1*(t(n) - t1) - 0.5*g*(t(n) - t1)^2;
```

```
end
v2 = v(n); h2 = h(n); t2 = t(n);
% Segment 3
while h(n) > 0 & n < 50000                          The third while loop.
  n = n + 1;
   t(n) = t(n - 1) + Dt;
   v(n) = vChute;
   h(n) = h2 + vChute*(t(n) - t2);
end
subplot(1,2,1)
plot(t,h,t2,h2,'o')
subplot(1,2,2)
plot(t,v,t2,v2,'o')
```

The accuracy of the results depends on the magnitude of the time increment Dt. An increment of 0.01s appears to give good results. The conditional expression in the while commands also includes a condition for n (if n is larger than 50,000 the loop stops). This is done as a precaution to avoid an infinite loop in case there is an error in the statements inside the loop. The plots generated by the program are shown below (axes labels and text were added to the plots using the plot editor).

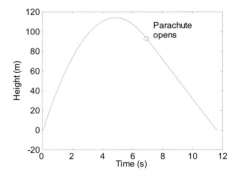
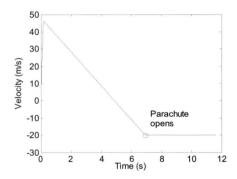

Note:

The problem can be solved and programmed in different ways. The solution shown here is one option. For example, instead of using while loops, the times that the parachute opens and the rocket hits the ground can be calculated first, and then for-end loops can be used instead of the while loop. If the times are determined first, it is possible also to use element-by-element calculations instead of loops.

Sample Problem 7-12: AC to DC convertor

A half-wave diode rectifier is an electrical circuit that converts AC voltage
to DC voltage. A rectifier circuit that
consists of an AC voltage source, a
diode, a capacitor, and a load (resistor) is shown in the figure. The voltage of the source is $v_s = v_0\sin(\omega t)$,
where $\omega = 2\pi f$, in which f is the frequency. The operation of the circuit is
illustrated in the diagram on the right
where the dashed line shows the
source voltage, and the solid line
shows the voltage across the resistor.
In the first cycle, the diode is on (con-
ducting current) from $t = 0$ until

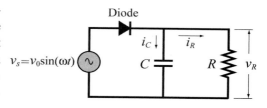

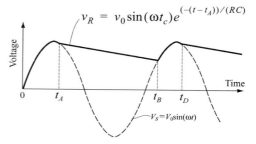

$t = t_A$. At this time the diode turns off and the power to the resistor is supplied by
the discharging capacitor. At $t = t_B$ the diode turns on again and continues to con-
duct current until $t = t_D$. The cycle continues as long as the voltage source is on.
In a simplified analysis of this circuit, the diode is assumed to be ideal and the
capacitor is assumed to have no charge initially (at $t = 0$). When the diode is on,
the resistor's voltage and current are given by:

$$v_R = v_0\sin(\omega t) \quad \text{and} \quad i_R = v_0\sin(\omega t)/R$$

The current in the capacitor is:

$$i_C = \omega C v_0\cos(\omega t)$$

When the diode is off, the voltage across the resistor is given by:

$$v_R = v_0\sin(\omega t_A)e^{(-(t-t_A))/(RC)}$$

The times when the diode switches off (t_A, t_D, and so on) are calculated from the
condition $i_R = -i_C$. The diode switches on again when the voltage of the source
reaches the voltage across the resistor (time t_B in the figure).

Write a MATLAB program that plots the voltage across the resistor v_R and
the voltage of the source v_s as a function of time for $0 \le t \le 70$ ms. The resistance
of the load is 1800 Ω, the voltage source $v_0 = 12$ V, and $f = 60$ Hz. To examine
the effect of the capacitor size on the voltage across the load, execute the program
twice; once with $C = 45$ μF, and once with $C = 10$ μF.

Solution

A program that solves the problem is listed below. The program has two parts.
One that calculates the voltage v_R when the diode is on, and the other when the

diode is off. The `switch` command is used for switching between the two parts. The calculations start with the diode on (the variable `state='on'`), and when $i_R - i_C \leq 0$ the value of `state` is changed to `'off'`, and the program switches to the commands that calculate v_R for this state. These calculations continue until $v_s \geq v_R$ when the program switches back to the equations that are valid when the diode is on.

```matlab
V0 = 12; C = 45e-6; R = 1800; f = 60;
Tf = 70e-3; w = 2*pi*f;
clear t VR Vs
t = 0:0.05e-3:Tf;
n = length(t);
state = 'on'
for i = 1:n
    Vs(i) = V0*sin(w*t(i));
    switch state
        case 'on'
        VR(i) = Vs(i);
        iR = Vs(i)/R;
        iC = w*C*V0*cos(w*t(i));
        sumI = iR + iC;
        if sumI <= 0
            state = 'off';
            tA = t(i);
        end
        case 'off'
        VR(i) = V0*sin(w*tA)*exp(-(t(i) - tA)/(R*C));
        if Vs(i) >= VR(i)
            state = 'on';
        end
    end
end
plot(t,Vs,':',t,VR,'k','linewidth',1)
xlabel('Time (s)'); ylabel('Voltage (V)')
```

Annotations:

- Assign `'on'` to the variable `state`.
- Calculate the voltage of the source at time t.
- Diode is on.
- Check if $i_R - i_C \leq 0$.
- If true, assign `'off'` to `state`.
- Assign value to t_A.
- Diode is off.
- Check if $v_s \geq v_R$.
- If true, assign `'on'` to the variable `state`.

The two plots generated by the program are shown below. One plot shows the result with $C = 45\ \mu F$, and the other with $C = 10\ \mu F$. It can be observed that with larger capacitor the DC voltage is smoother (smaller ripple in the wave).

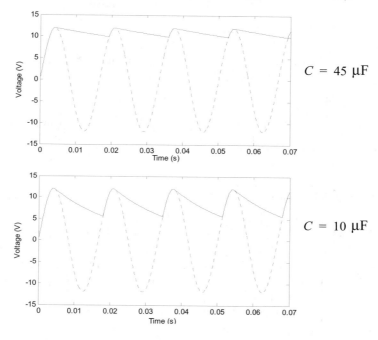

$C = 45\ \mu F$

$C = 10\ \mu F$

7.8 PROBLEMS

1. Evaluate the following expressions without using MATLAB. Check the answer with MATLAB.

   a) $5 <= 8 - 3$

   b) $y = 7 < 3 - 1 + 6 > 2$

   c) $y = (7 < 3) - 1 + (6 > 2)$

   d) $y = 2 \times 4 + 5 == 7 + \dfrac{20}{4}$

2. Given: $a = 10$, $b = 6$. Evaluate the following expressions without using MATLAB. Check the answer with MATLAB.

   a) $y = a >= b$

   b) $y = a - b <= \dfrac{b}{2}$

   c) $y = a - \left(b <= \dfrac{b}{2}\right)$

3. Given: $v = [4\ -2\ -1\ 5\ 0\ 1\ -3\ 8\ 2]$ and $w = [0\ 2\ 1\ -1\ 0\ -2\ 4\ 3\ 2]$. Evaluate the following expressions without using MATLAB. Check the answer with MATLAB.

   a) $v \ >= \ w$
   b) $w \sim= \ v$

4. Use the vectors v and w from the previous problem. Use relational operators to create a vector y that is made up from the elements of w that are greater than the elements of v.

5. Evaluate the following expressions without using MATLAB. Check the answer with MATLAB.

   a)   5&–2
   b)   8-2|6+5&~2
   c)   ~(4&0)+8*~(4|0)

6. The maximum daily temperature (in °F) for New York City and Anchorage, Alaska during the month of January, 2001 are given in the vectors below (data from the U.S. National Oceanic Atmospheric Administration).
   TNY = [31 26 30 33 33 39 41 41 34 33 45 42 36 39 37 45 43 36 41 37 32 32 35 42 38 33 40 37 36 51 50]
   TANC = [37 24 28 25 21 28 46 37 36 20 24 31 34 40 43 36 34 41 42 35 38 36 35 33 42 42 37 26 20 25 31]
   Write a program in a script file to answer the following:
   a)   Calculate the average temperature for the month in each city.
   b)   How many days was the temperature below the average in each city?
   c)   How many days, and which dates in the month, was the temperature in Anchorage higher then the temperature in New York?
   d)   How many days, and which dates in the month, was the temperature the same in both cities?
   e)   How many days, and which dates in the month, was the temperature in both cities above freezing (above 32 °F)?

7. Use MATLAB in the two different ways, described below, to plot the function:

$$f(x) = \begin{cases} 4e^{x+2} & \text{for} & -6 \le x \le -2 \\ x^2 & \text{for} & -2 \le x \le 2.5 \\ (x+6.5)^{1/3} & \text{for} & 2.5 \le x \le 6 \end{cases}$$

   a)   Write a program in a script file, using conditional statements and loops.
   b)   Create a user-defined function for $f(x)$, and then use the function in a script file to make the plot.

8. Write a program in a script file that determines the real roots of a quadratic equation $ax^2 + bx + c = 0$. Name the file quadroots. When the file runs it asks the user to enter the values of the constants a, b, and c. To calculate the roots of the equation the program calculates the discriminant D given by:

$$D = b^2 - 4ac$$

If $D > 0$ the program displays a message: "The equation has two roots," and the roots are displayed in the next line.
If $D = 0$ the program displays a message: "The equation has one root," and the root is displayed in the next line.
If $D < 0$ the program displays a message: "The equation has no real roots."
Run the script file in the Command Window three times to obtain solution to the following three equations:

a) $2x^2 + 8x - 3 = 0$
b) $15x^2 + 10x + 5 = 0$
c) $18x^2 + 12x + 2 = 0$.

9. Use loops to create a 4×7 matrix in which the value of each element is the sum of its indices (the row number and column number of the element). For example, the value of element A(2,5) is 7.

10. Use loops and conditional statements to create a 5×8 matrix in which the value of each element is equal to the square root of the sum of the element's indices unless the element is in an even-numbered column or row. The value of an element in an even-numbered column or row is equal to the sum of the element's indices squared. (The indices of an element in a matrix are the row number and column number of the element.)

11 Write a program (using a loop) that determines the sum of the first m terms of the series:

$$\sum_{n=0}^{m} (-1)^n \frac{1}{2n+1} \quad (n = 0, 1, 2,..., m)$$

Run the program with $m = 10$, and $m = 500$. Compare the result with $\pi/4$. This series which is called the Leibniz series converges to $\pi/4$.

12. A vector is given by: $x = [15 \ -6 \ 0 \ 8 \ -2 \ 5 \ 4 \ -10 \ 0.5 \ 3]$. Using conditional statements and loops write a program that determines the sum of the positive elements in the vector.

13. Write a program in a script file that finds the smallest odd integer that is also divisible by 3 and whose cube is greater than 4,000. Use a loop in the program. The loop should start from 1 and stop when the number is found. The

program prints a message: "The required number is:" and then prints the number.

14. Write a user-defined function that sorts the elements of a vector (of any length) from the largest to the smallest. For the function name and arguments use y = downsort (x). The input to the function is a vector x of any length, and the output y is a vector in which the elements of x are arranged in descending order. Do not use the MATLAB sort function. Test your function on a vector with 14 numbers (integers) randomly distributed between -30 and 30. Use the MATLAB rand function to generate the initial vector.

15. Write a user-defined function that sorts the elements of a matrix. For the function name and arguments use B = matrixsort (A), where A is any size matrix and B is a matrix of the same size with the elements of A rearranged in an ascending order row after row where the (1,1) element is the smallest, and the (m,n) element is the largest.

    Test your function on a 4×7 matrix with elements (integers) randomly distributed between -30 and 30. Use the MATLAB rand function to generate the initial matrix.

16. Write a program in a script file that calculates the cost of mailing a package according to the following price schedule:

Type of service	Weight 0–2 lb.	Weight 2–10 lb.	Weight 10–50 lb.
Ground	$1.50	$1.50 + $0.50 for each pound or fraction of a pound above 2 lb.	$5.50 + $0.30 for each pound or fraction of a pound above 10 lb.
Air	$3.00	$3.00 + $0.90 for each pound or fraction of a pound above 2 lb.	$10.20 + $0.60 for each pound or fraction of a pound above 10 lb.
Over-night	$18	$18.00 + $6.00 for each pound or fraction of a pound above 2 lb.	No overnight service for packages above 10 lb.

The program asks the user to enter the weight and the type of service. The program then displays the cost. If a weight larger than 50 lb is entered for ground or air service, a message "Ground (or Air) service is not available for packages that weigh more than 50 lb" is displayed. If a weight larger than 10 lb is entered for overnight service a message "Overnight service is not available for packages that weigh more than 10 lb" is displayed. Run the program and enter 0.5, 6.3, 20, and 50.4 lb for Ground and Air service, and 2, 8.1 and 13 lb for Overnight service.

17. A vector x is given by: $x = [1{:}50]$. Write a program in a script file that removes all the vector's elements that are divisible by 3, 4, or 5 and displays the new vector.

18. Write a user-defined function that determines the polar coordinates of a point from the Cartesian coordinates in a two-dimensional plane. For the function name and arguments use `[theta radius]=CartesianTo-Polar(x,y)`. The input arguments are the x and y coordinates of the point, and the output arguments are the angle θ and the radial distance to the point. The angle θ is in degrees and is measured relative to the positive x axis, such that it is a positive number in quadrants I, II, and III, and a negative number in the IV quadrant. Use the function to determine the polar coordinates of points $(15, 3)$, $(-7, 12)$, $(-17, -9)$, and $(10, -6.5)$.

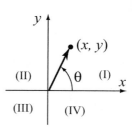

19. A cylindrical, vertical fuel tank has hemispheric end caps as shown. The radius of the cylinder and the caps is $r = 40$ cm, and the length of the cylindrical part is 1.2 m.

    Write a user-defined function (for the function name and arguments use `V = Vfuel(h)`) that gives the volume of the fuel in the tank as a function of the height h. Use the function to make a plot of the volume as a function of h for $0 \le h \le 2$ m.

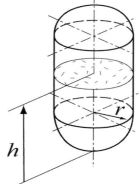

20. The velocity, as a function of time of a particle that moves along a straight line, is shown on the right and given in the equations below.

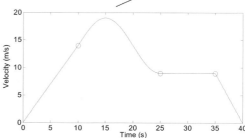

$$v(t) = \begin{cases} 1.4t & \text{for} & 0 \le t \le 10 \text{ s} \\ 14 + 5\sin\left(\dfrac{\pi}{10}(t-10)\right) & \text{for} & 10 \le t \le 25 \text{ s} \\ 9 & \text{for} & 25 \le t \le 35 \text{ s} \\ 9 - \dfrac{9}{5}(t-35) & \text{for} & 35 \le t \le 40 \text{ s} \end{cases}$$

Write two user-defined functions: One that calculates the velocity of the particle at time t (for the function name and arguments use v = velocity(t)),
and the other that calculates the acceleration of the particle at time t (for the function name and arguments use a = acceleration(t)). In a script file, write a
program that creates plots of the velocity and acceleration as functions of time
(two plots on the same page). In the program, first create a vector t, $0 \leq t \leq 40$ s,
and then use the functions velocity and acceleration to create vectors of
velocity and acceleration that are used for the plots.

21. A scale is made of a tray attached to springs as shown in the figure. When an
object is placed on the tray, the tray moves down and the weight of the object can
be determined from the displacement of the tray. Initially, only the two outside
springs support the weight. If the object is heavy enough, however, the tray makes
contact with a third spring in the middle.

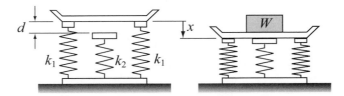

$k_1 = 800$ N/m, $k_2 = 1700$ N/m, $d = 20$ mm.

Write a user-defined function that calculates the weight W of the object for a given
displacement x of the tray. For the function name and arguments use
W = scale(x).

a) Using the function in the Command Window, determine the weight of two
objects for tray displacements of 1.5 and 3.1 cm.

b) Write a program in a script file that plots the weight as a function of displacement for $0 \leq x \leq 4$ cm.

Chapter 8

Polynomials, Curve Fitting, and Interpolation

Polynomials are mathematical expressions that are frequently used for problem solving and modeling in science and engineering. In many cases an equation that is written in the process of solving a problem is a polynomial, and the solution of the problem is the zero of the polynomial. MATLAB has a wide selection of functions that are specifically designed for handling polynomials. How to use polynomials in MATLAB is described in Section 8.1.

Curve fitting is a process of finding a function that can be used to model data. The function does not necessarily pass through any of the points, but models the data with the smallest possible error. There are no limitations to the type of the equations that can be used for curve fitting. Often, however, polynomial, exponential, and power functions are used. In MATLAB curve fitting can be done by writing a program, or by interactively analyzing data that is displayed in the Figure Window. Section 8.2 describes how to use MATLAB programming for curve fitting with polynomials and other functions. Section 8.4 describes the basic fitting interface which is used for interactive curve fitting and interpolation.

Interpolation is the process of estimating values between data points. The simplest interpolation is done by drawing a straight line between the points. In a more sophisticated interpolation, data from additional points is used. How to interpolate with MATLAB is discussed in Sections 8.3 and 8.4.

8.1 POLYNOMIALS

Polynomials are functions that have the form:

$$f(x) = a_n x^n + a_{n-1} x^{n-1} + \ldots + a_1 x + a_0$$

The coefficients $a_n, a_{n-1}, \ldots, a_1, a_0$ are real numbers, and n, which is a nonnega-

tive integer, is the degree, or order, of the polynomial.
Examples of polynomials are:

$$f(x) = 5x^5 + 6x^2 + 7x + 3 \qquad \text{polynomial of degree 5.}$$
$$f(x) = 2x^2 - 4x + 10 \qquad \text{polynomial of degree 2.}$$
$$f(x) = 11x - 5 \qquad \text{polynomial of degree 1.}$$

A constant (e.g. $f(x) = 6$) is a polynomial of degree 0.

In MATLAB, polynomials are represented by a row vector in which the elements are the coefficients $a_n, a_{n-1}, ..., a_1, a_0$. The first element is the coefficient of the x with the highest power. The vector has to include all the coefficients, including the ones that are equal to 0. For example:

Polynomial	**MATLAB representation**
$8x + 5$	p = [8 5]
$2x^2 - 4x + 10$	d = [2 –4 10]
$6x^2 - 150$, MATLAB form: $6x^2 + 0x - 150$	h = [6 0 –150]
$5x^5 + 6x^2 - 7x$, MATLAB form:	c = [5 0 0 6 –7 0]
$\qquad 5x^5 + 0x^4 + 0x^3 + 6x^2 - 7x + 0$	

8.1.1 Value of a Polynomial

The value of a polynomial at a point x can be calculated with the function `polyval` that has the form:

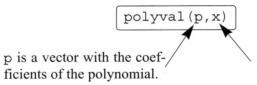

p is a vector with the coefficients of the polynomial.

x is a number, or a variable that has an assigned value, or a computable expression.

x can also be a vector or a matrix. In such a case the polynomial is calculated for each element (element-by-element), and the answer is a vector, or a matrix, with the corresponding values of the polynomial.

Sample Problem 8-1: Calculating polynomials with MATLAB

For the polynomial: $f(x) = x^5 - 12.1x^4 + 40.59x^3 - 17.015x^2 - 71.95x + 35.88$

a)   Calculate $f(9)$.

b)   Plot the polynomial for $-1.5 \le x \le 6.7$.

Solution

The problem is solved in the Command Window.

a)   The coefficients of the polynomials are assigned to vector p. The function `ployval` is then used to calculate the value at $x = 9$.

```
>> p = [1 -12.1 40.59 -17.015 -71.95 35.88];
>> polyval(p,9)

ans =
  7.2611e+003
```

b) To plot the polynomial, a vector x is first defined with elements ranging from -1.5 to 6.7. Then a vector y is created with the values of the polynomial for every element of x. Finally, a plot of y vs. x is made.

```
>> x = -1.5:0.1:6.7;
>> y = polyval(p,x);          Calculating the value of the polyno-
>> plot(x,y)                  mial for each element of the vector x.
```

The plot created by MATLAB is (axes labels were added with the Plot Editor):

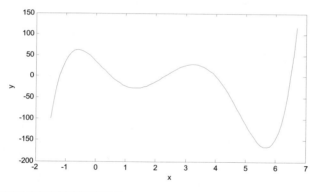

8.1.2 Roots of a Polynomial

Roots of a polynomial are the values of the argument for which the value of the polynomial is equal to zero. For example, the roots of the polynomial $f(x) = x^2 - 2x - 3$ are the values of x for which $x^2 - 2x - 3 = 0$, which are $x = -1$, and $x = 3$.

MATLAB has a function, called `roots`, that determines the root, or roots, of a polynomial. The form of the function is:

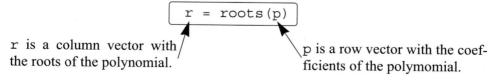

r is a column vector with the roots of the polynomial.

p is a row vector with the coefficients of the polymomial.

For example, the roots of the polynomial in Sample Problem 8-1 can be determined by:

```
>> p = [1 -12.1 40.59 -17.015 -71.95 35.88];
```

```
>> r = roots(p)
```

```
r =
    6.5000
    4.0000
    2.3000
   -1.2000
    0.5000
```

> When the roots are known, the polynomial can actually be written as:
> $$f(x) = (x + 1.2)(x - 0.5)(x - 2.3)(x - 4)(x - 6.5)$$

The `roots` command is very useful for finding the roots of a quadratic equation. For example, to find the roots of $f(x) = 4x^2 + 10x - 8$ type:

```
>> roots([4 10 -8])
```

```
ans =
   -3.1375
    0.6375
```

When the roots of a polynomial are known, the `poly` command can be used for determining the coefficients of the polynomial. The form of the `poly` command is:

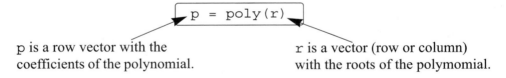

p is a row vector with the coefficients of the polynomial.

r is a vector (row or column) with the roots of the polymomial.

For example, the coefficients of the polynomial in Sample Problem 8-1 can be obtained from the roots of the polynomial (see above) by:

```
>> r = [6.5 4 2.3 -1.2 0.5];
```

```
>> p = poly(r)
```

```
p =
    1.0000 -12.1000   40.5900  -17.0150  -71.9500   35.8800
```

8.1.3 Addition, Multiplication, and Division of Polynomials

Addition:

Two polynomials can be added (or subtracted) by adding the vectors of the coefficients. If the polynomials are not of the same order (which means that the vectors of the coefficients are not of the same length), the shorter vector has to be modified to be of the same length as the longer vector by adding zeros (called padding) in front. For example, the polynomials:

$f_1(x) = 3x^6 + 15x^5 - 10x^3 - 3x^2 + 15x - 40$ and $f_2(x) = 3x^3 - 2x - 6$ can be added by:

```
>> p1 = [3 15 0 -10 -3 15 -40];
>> p2 = [3 0 -2 -6];
>> p = p1 + [0 0 0 p2]
p =
    3   15   0   -7   -3   13   -46
```

Three 0's are added in front of p2, since the order of p1 is 6 and the order of p2 is 3.

Multiplication:

Two polynomials can be multiplied with the MATLAB built-in function `conv` which has the form:

$$c = conv(a,b)$$

c is a vector of the coefficients of the polynomial that is the product of the multiplication.

a and b are the vectors of the coefficients of the polynomials that are being multiplied.

- The two polynomials do not have to be of the same order.

- Multiplication of three or more polynomials is done by using the `conv` function repeatedly.

For example, multiplication of the polynomials $f_1(x)$ and $f_2(x)$ above gives:

```
>> pm = conv(p1,p2)
pm =
    9   45   -6   -78   -99   65   -54   -12   -10   240
```

which means that the answer is:

$9x^9 + 45x^8 - 6x^7 - 78x^6 - 99x^5 + 65x^4 - 54x^3 - 12x^2 - 10x + 240$

Division:

A polynomial can be divided by another polynomial with the MATLAB built-in function `deconv` which has the form:

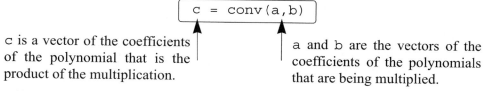

$$[q,r] = deconv(u,v)$$

q is a vector with the coefficients of the quotient polynomial.
r is a vector with the coefficients of the remainder polynomial.

u is a vector with the coefficients of the numerator polynomial.
v is a vector with the coefficients of the denominator polynomial.

For example, dividing $2x^3 + 9x^2 + 7x - 6$ by $x + 3$ is done by:

```
>> u = [2 9 7 -6];
>> v = [1 3];
>> [a b] = deconv(u,v)
a =
    2   3   -2
b =
    0   0   0   0
```

The answer is: $2x^2 + 3x - 2$

Remainder is zero.

An example of division that gives a remainder is $2x^6 - 13x^5 + 75x^3 + 2x^2 - 60$ divided by $x^2 - 5$:

```
>> w = [2 -13 0 75 2 0 -60];
>> z = [1 0 -5];
>> [g h] = deconv(w,z)
g =
    2  -13  10  10  52
h =
    0   0   0   0   0  50  200
```

The quotient is: $2x^4 - 13x^3 + 10x^2 + 10x + 52$

The remainder is: $50x - 200$

The answer is: $2x^4 - 13x^3 + 10x^2 + 10x + 52 + \dfrac{50x + 200}{x^2 - 5}$.

8.1.4 Derivatives of Polynomials

The built-in function `polyder` can be used to calculate the derivative of a single polynomial, a product of two polynomials, and a quotient of two polynomials, as shown in the following three commands.

`k = polyder(p)`	Derivative of a single polynomial. p is a vector with the coefficients of the polynomial. k is a vector with the coefficients of the polynomial that is the derivative.
`k = polyder(a,b)`	Derivative of a product of two polynomials. a and b are vectors with the coefficients of the polynomials that are multiplied. k is a vector with the coefficients of the polynomial that is the derivative of the product.
`[n d] = polyder(u,v)`	Derivative of a quotient of two polynomials. u and v are vectors with the coefficients of the numerator and denominator polynomials. n and d are vectors with the coefficients of the numerator and denominator polynomials in the quotient that is the derivative.

The only difference between the last two commands is in the number of output arguments. With two output arguments MATLAB calculates the derivative of quotient of two polynomials. With one output argument the derivative is of the product.

For example, if $f_1(x) = 3x^2 - 2x + 4$, and $f_2(x) = x^2 + 5$, the derivatives of $3x^2 - 2x + 4$, $(3x^2 - 2x + 4)(x^2 + 5)$, and $\dfrac{3x^2 - 2x + 4}{x^2 + 5}$ can be determined by:

```
>> f1 = [3 -2 4]; f2 = [1 0 5];
```
Creating the vectors coefficients of f_1 and f_2.
```
>> k = polyder(f1)
k =
    6  -2
```
The derivative of f_1 is: $6x - 2$.
```
>> d = polyder(f1,f2)
d =
    12  -6  38  -10
```
The derivative of f_1*f_2 is: $12x^3 - 6x^2 + 38x - 10$.
```
>> [n d] = polyder(f1,f2)
n =
    2  22  -10
d =
    1  0  10  0  25
```
The derivative of $\dfrac{3x^2 - 2x + 4}{x^2 + 5}$ is: $\dfrac{2x^2 + 22x - 10}{x^4 + 10x^2 + 25}$.

8.2 CURVE FITTING

Curve fitting, also called regression analysis, is a process of fitting a function to a set of data points. The function can then be used as a mathematical model of the data. Since there are many types of functions (linear, polynomial, power, exponential, etc.) curve fitting can be a complicated process. Many times there is some idea of the type of function that might fit the given data and the need is only to determine the coefficients of the function. In other situations, where nothing is known about the data, it is possible to make different types of plots which provide information about possible forms of functions that might give a good fit to the data. This section describes some of the basic techniques for curve fitting, and the tools that MATLAB has for this purpose.

8.2.1 Curve Fitting with Polynomials, the `polyfit` Function

Polynomials can be used to fit data points in two ways. In one, the polynomial passes through all the data points, and in the other the polynomial does not necessarily pass through any of the points, but overall gives a good approximation of the data. The two options are described below.

Polynomials that pass through all the points:

When n points (x_i, y_i) are given, it is possible to write a polynomial of degree $n - 1$

that passes through all the points. For example, if two points are given it is possible to write a linear equation in the form of $y = mx + b$ that passes through the points. With three points the equation has the form of $y = ax^2 + bx + c$. With n points the polynomial has the form: $a_{n-1}x^{n-1} + a_{n-2}x^{n-2} + \ldots + a_1x + a_0$. The coefficients of the polynomial are determined by substituting each point in the polynomial, and then solving the n equations for the coefficients. As will be shown later in this section, polynomials of high degree might give a large error if they are used to estimate values between data points.

Polynomials that do not necessarily pass through any of the points:

When n points are given, it is possible to write a polynomial of degree less than $n-1$ that does not necessarily pass through any of the points, but overall approximates the data. The most common method of finding the best fit to data points is the method of least squares. In this method the coefficients of the polynomial are determined by minimizing the sum of the squares of the residuals at all the data points. The residual at each point is defined as the difference between the value of the polynomial and the value of the data. For example, consider the case of finding the equation of a straight line that best fits four data points as shown in Figure 8-1. The points are (x_1, y_1), (x_2, y_2), (x_3, y_3), (x_4, y_4), and the polynomial of the

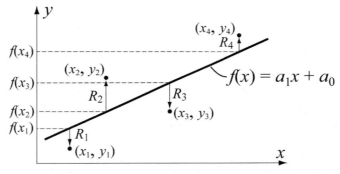

Figure 8-1: Least square fitting of first degree polynomial to four points.

first degree can be written by: $f(x) = a_1x + a_0$. The residual, R_i at each point, is the difference between the value of the function at x_i and y_i, $R_i = f(x_i) - y_i$. An equation for the sum of the squares of the residuals R_i of all the points is given by:

$$R = [f(x_1) - y_1]^2 + [f(x_2) - y_2]^2 + [f(x_3) - y_3]^2 + [f(x_4) - y_4]^2$$

or, after substituting the equation of the polynomial at each point, by:

$$R = [a_1x_1 + a_0 - y_1]^2 + [a_1x_2 + a_0 - y_2]^2 + [a_1x_3 + a_0 - y_3]^2 + [a_1x_4 + a_0 - y_4]^2$$

At this stage R is a function of a_1 and a_0. The minimum of R can be determined by taking the partial derivative of R with respect to a_1 and a_0 (2 equations) and equating them to zero.

$$\frac{\partial R}{\partial a_1} = 0 \quad \text{and} \quad \frac{\partial R}{\partial a_0} = 0$$

This results in a system of two equations with two unknowns, a_1 and a_0. The solution of these equations gives the values of the coefficients of the polynomial that best fits the data. The same procedure can be followed with more points and higher order polynomials. More details on the least squares method can be found in books on numerical analysis.

Curve fitting with polynomials is done in MATLAB with the `polyfit` function, which uses the least squares method. The basic form of the `polyfit` function is:

p is the vector of the coefficients of the polynomial that fits the data.

x is a vector with the horizontal coordinate of the data points (independent variable).
y is a vector with the vertical coordinate of the data points (dependent variable).
n is the degree of the polynomial.

For the same set of m points, the `polyfit` function can be used to fit polynomials of any order up to $m - 1$. If $n = 1$ the polynomial is a straight line, if $n = 2$ the polynomial is a parabola, and so on. The polynomial passes through all the points if $n = m - 1$ (the order of the polynomial is one less than the number of points). It should be pointed out here that a polynomial that passes through all the points, or polynomials with higher order, do not necessarily give a better fit overall. High order polynomials can sometimes deviate significantly between the data points.

Figure 8-2 shows how polynomials of different degrees fit the same set of data points. A set of seven points is given by: (0.9, 0.9), (1.5, 1.5), (3, 2.5),

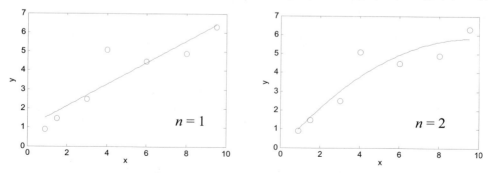

Figure 8-2: Fitting data with polynomials of different order.

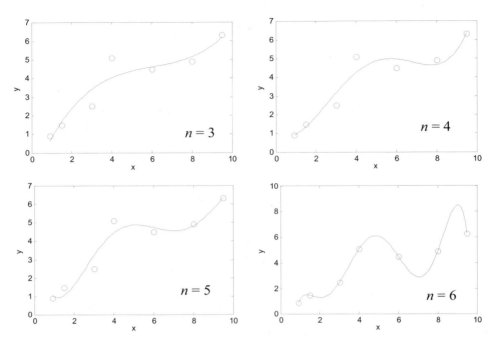

Figure 8-2: Fitting data with polynomials of different order. (Continued)

(4, 5.1), (6, 4.5), (8, 4.9), and (9.5, 6.3). The points are fitted using the `polyfit` function with polynomials of degrees one through six. Each plot in Figure 8-2, shows the same data points, marked with circles, and a curve-fitted line that corresponds to a polynomial with the specified degree. It can be seen that the polynomial with $n = 1$ is a straight line, and with $n = 2$ is a slightly curved line. As the degree of the polynomial increases the line has more bends such that it passes closer to more points. When $n = 6$, which is one less than the number of points, the line passes through all the points. However, between some of the points, the line deviates significantly from the trend of the data.

The script file used to generate one of the plots in Figure 8-2 (the polynomial with $n = 3$) is shown below. Note that in order to plot the polynomial (the

```
x = [0.9 1.5 3 4 6 8 9.5];          Create vectors x and y with the
y = [0.9 1.5 2.5 5.1 4.5 4.9 6.3];  coordinates of the data points.
p = polyfit(x,y,3)        Create a vector p using the polyfit function.
xp = 0.9:0.1:9.5;         Create a vector xp to be used for plotting the polynomial.
yp = polyval(p,xp);       Create a vector yp with values of the polynomial at each xp.
plot(x,y,'o',xp,yp)       A plot of the 7 points and the polynomial.
xlabel('x'); ylabel('y')
```

line) a new vector `xp` with small spacing is created. This vector is then used with the function `polyval` to create a vector `yp` with the value of the polynomial for each element of `xp`.

When the script file is executed, the following vector `p` is displayed in the Command Window.

```
p =
    0.0220  -0.4005   2.6138  -1.4158
```

This means that the polynomial of the third degree in Figure 8-2 has the form:
$0.022x^3 - 0.4005x^2 + 2.6138x - 1.4148$.

8.2.2 Curve Fitting with Functions Other than Polynomials

Many situations in science and engineering require fitting functions that are not polynomials to given data. Theoretically, any function can be used to model data within some range. For a particular data set, however, some functions provide a better fit than others. In addition, determining the best fit coefficients of some functions can be more difficult than others. This section covers curve fitting with power, exponential, logarithmic, and reciprocal functions, which are commonly used. The forms of these functions are:

$$y = bx^m \qquad \text{(power function)}$$

$$y = be^{mx} \quad \text{or} \quad y = b10^{mx} \qquad \text{(exponential function)}$$

$$y = m\ln(x) + b \quad \text{or} \quad y = m\log(x) + b \quad \text{(logarithmic function)}$$

$$y = \frac{1}{mx + b} \qquad \text{(reciprocal function)}$$

All of these functions can easily be fitted to given data with the `polyfit` function. This is done by rewriting the functions in a form that can be fitted with a linear polynomial ($n = 1$) which has the form:

$$y = mx + b$$

The logarithmic function is already in this form, and the power, exponential and reciprocal equations can be rewritten as:

$$\ln(y) = m\ln(x) + \ln b \qquad \text{(power function)}$$

$$\ln(y) = mx + \ln(b) \quad \text{or} \quad \log(y) = mx + \log(b) \qquad \text{(exponential function)}$$

$$\frac{1}{y} = mx + b \qquad \text{(reciprocal equation)}$$

These equations describe a linear relationship between $\ln(y)$ and $\ln(x)$ for the power function, between $\ln(y)$ and x for the exponential function, between y and $\ln(x)$ or $\log(x)$ for the logarithmic function, and between $1/y$ and x for the reciprocal function. This means that the `polyfit(x,y,1)` function can be used to determine the best fit constants m and b for best fit if, instead of `x` and `y`, the following arguments are used.

Function		polyfit function form
power	$y = bx^m$	p=polyfit(log(x),log(y),1)
exponential	$y = be^{mx}$ or	p=polyfit(x,log(y),1) or
	$y = b10^{mx}$	p=polyfit(x,log10(y),1)
logarithmic	$y = m\ln(x) + b$ or	p=polyfit(log(x),y,1) or
	$y = m\log(x) + b$	p=polyfit(log10(x),y,1)
reciprocal	$y = \dfrac{1}{mx + b}$	p=polyfit(x,1./y,1)

The result of the polyfit function is assigned to p, which is a two element vector. The first element, p(1), is the constant m, and the second element, p(2), is b for the logarithmic and reciprocal functions, $\ln(b)$ or $\log(b)$ for the exponential function, and $\ln(b)$ for the power function ($b = e^{p(2)}$ or $b = 10^{p(2)}$ for the exponential function and $b = e^{p(2)}$ for the power function).

For given data it is possible to foresee, to some extent, which of the functions has the potential for providing a good fit. This is done by plotting the data using different combinations of linear and logarithmic axes. If the data points in one of the plots appear to fit a straight line, the corresponding function can provide a good fit according to the list below.

x axis	**y axis**	**Function**
linear	linear	linear $y = mx + b$
logarithmic	logarithmic	power $y = bx^m$
linear	logarithmic	exponential $y = be^{mx}$ or $y = b10^{mx}$
logarithmic	linear	logarithmic $y = m\ln(x) + b$ or $y = m\log(x) + b$
linear	linear (plot $1/y$)	reciprocal $y = \dfrac{1}{mx + b}$

Other considerations when choosing a function are:

- Exponential functions can not pass through the origin.
- Exponential functions can only fit data with all positive y's, or all negative y's.
- Logarithmic functions can not model $x = 0$, or negative values of x.
- For the power function $y = 0$ when $x = 0$.
- The reciprocal equation can not model $y = 0$.

The following example illustrates the process of fitting a function to a set of data points.

Sample Problem 8-2:   Fitting an equation to data points

The following data points are given. Determine a function $w = f(t)$ (t is the independent variable, w is the dependent variable) with a form discussed in this section that best fits the data.

t	0.0	0.5	1.0	1.5	2.0	2.5	3.0	3.5	4.0	4.5	5.0
w	6.00	4.83	3.70	3.15	2.41	1.83	1.49	1.21	0.96	0.73	0.64

Solution

The data is first plotted with linear scales on both axes. The figure, shown on the right, indicates that a linear function will not give the best fit since the points do not appear to line up along a straight line. From the other possible functions, the logarithmic function is excluded since in the first point $t = 0$, and the power function is excluded since at $t = 0$, $w \neq 0$. To check if 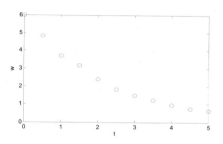 the other two functions (exponential and reciprocal) might give a better fit, two additional plots, shown below, are made. The plot on the left has a log scale on the vertical axis, and linear horizontal axis. In the plot on the right both axes have linear scales, and the quantity $1/w$ is plotted on the vertical axis.

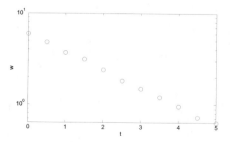
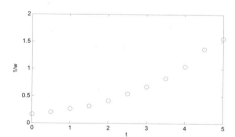

In the left figure the data points appear to line up along a straight line. This indicates that an exponential function in the form $y = be^{mx}$ can give a good fit to the data. A program in a script file that determines the constants b and m, and plots the data points and the function is given below.

```
t = 0:0.5:5;        Create vectors t and w with the coordinates of the data points.
w = [6 4.83 3.7 3.15 2.41 1.83 1.49 1.21 0.96 0.73 0.64];
p = polyfit(t,log(w),1);        Use the polyfit function with t and log(w).
```

```
m = p(1)
b = exp(p(2))                         Determine the coefficient b.
tm = 0:0.1:5;              Create a vector tm to be used for plotting the polynomial.
wm = b*exp(m*tm);            Calculate the function value at each element of tm.
plot(t,w,'o',tm,wm)              Plot the data points and the function.
```

When the program is executed, the values of the constants m and b are displayed in the Command Window.

```
m =
  -0.4580
b =
   5.9889
```

The plot, generated by the program, that shows the data points and the function is (axes labels were added with the Plot Editor):

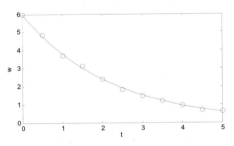

It should be pointed out here that in addition to the power, exponential, logarithmic, and reciprocal functions that are discussed in this section, many other functions can be written in a form suitable for curve fitting with the `polyfit` function. One example where a function of the form $y = e^{(a_2 x^2 + a_1 x + a_0)}$ is fitted to data points using the `polyfit` function with a third order polynomial is described in Sample Problem 8-7.

8.3 INTERPOLATION

Interpolation is estimation of values between data points. MATLAB has interpolation functions that are based on polynomials, which are described in this section, and on Fourier transformation, which are outside the scope of this book. In one-dimensional interpolation each point has one independent variable (x) and one dependent variable (y). In two-dimensional interpolation each point has two independent variables (x and y) and one dependent variable (z).

One-dimensional interpolation:

If only two data points exist, the points can be connected with a straight line and a linear equation (polynomial of first order) can be used to estimate values between the points. As was discussed in the previous section, if three (or four) data points exist, a second (or a third) order polynomial can be determined to pass through the points and then be used to estimate values between the points. As the number of points increases, a higher order polynomial is required for the polynomial to pass through all the points. Such a polynomial, however, will not necessarily give a good approximation of the values between the points. This is illustrated in Figure 8-2 with $n = 6$.

A more accurate interpolation can be obtained if instead of considering all the points in the data set (by using one polynomial that passes through all the points), only a few data points in the neighborhood where the interpolation is needed are considered. In this method, called spline interpolation, many low order polynomials are used, where each is valid only in a small domain of the data set.

The simplest method of spline interpolation is called linear spline interpolation. In this method, shown on the right, every two adjacent points are connected with a straight line (a polynomial of first degree). The equation of a straight line that passes through two adjacent points (x_i, y_i) and (x_{i+1}, y_{j+1}) and can be used to calculate the value of y for any x between the points is given by:

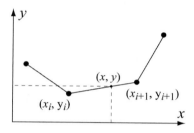

$$y = \frac{y_{i+1} - y_i}{x_{i+1} - x_i} x + \frac{y_i x_{i+1} - y_{i+1} x_i}{x_{i+1} - x_i}$$

In a linear interpolation the line between two data points has a constant slope, and there is a change in the slope at every point. A smoother interpolation curve can be obtained by using quadratic, or cubic polynomials. In these methods, called quadratic splines or cubic splines, a second, or third order polynomial is used to interpolate between every two points. The coefficients of the polynomial are determined by using data from additional points that are adjacent to the two data points. The theoretical background for the determination of the constants of the polynomials is beyond the scope of this book and can be found in books on numerical analysis.

One-dimensional interpolation in MATLAB is done with the `interp1` (last character is the number one) function, which has the form:

$$\boxed{\text{yi} = \text{interp1(x,y,xi,'method')}}$$

yi is the interpolated value.

x is a vector with the horizontal coordinate of the input data points (independent variable).
y is a vector with the vertical coordinate of the input data points (dependent variable).
xi is the horizontal coordinate of the interpolation point (independent variable).

Method of interpolation, typed as a string (optional).

- The vector x must be monotonic (the elements in ascending or descending order).

- xi can be a scalar (interpolation of one point) or a vector (interpolation of many points). Respectively, yi is a scalar or a vector with the corresponding interpolated values.

- MATLAB can do the interpolation using one of several methods that can be specified. These methods include:

  'nearest'   returns the value of the data point that is nearest to the interpolated point.
  'linear'   uses linear spline interpolation.
  'spline'   uses cubic spline interpolation.
  'pchip'   uses piecewise cubic Hermite interpolation, also called 'cubic'

- When the 'nearest' and the 'linear' methods are used, the value(s) of xi must be within the domain of x. If the 'spline' or the 'pchip' methods are used, xi can have values outside the domain of x and the function interp1 performs extrapolation.

- The 'spline' method can give large errors if the input data points are non-uniform such that some points are much closer together than others.

- Specification of the method is optional. If no method is specified the default is 'linear'.

Sample Problem 8-3: Interpolation

The following data points, which are points of the function $f(x) = 1.5^x \cos(2x)$, are given. Use linear, spline, and pchip interpolation methods to calculate the value of y between the points. Make a figure for each of the interpolation methods. In the figure show the points, a plot of the function, and a curve that corresponds to the interpolation method.

x	0	1	2	3	4	5
y	1.0	−0.6242	−1.4707	3.2406	−0.7366	−6.3717

Solution

The following is a program written in a script file that solves the problem:

```
x = 0:1.0:5;                    Create vectors x and y with coordinates of the data points.
y = [1.0 -0.6242 -1.4707 3.2406 -0.7366 -6.3717];
xi = 0:0.1:5;                   Create vector xi with points for interpolation.
yilin = interp1(x,y,xi,'linear');   Calculate y points from linear interpolation.
yispl = interp1(x,y,xi,'spline');   Calculate y points from spline interpolation.
yipch = interp1(x,y,xi,'pchip');    Calculate y points from pchip interpolation.
yfun = 1.5.^xi.*cos(2*xi);      Calculate y points from the function.
subplot(1,3,1)
plot(x,y,'o',xi,yfun,xi,yilin,'--');
subplot(1,3,2)
plot(x,y,'o',xi,yfun,xi,yispl,'--');
subplot(1,3,3)
plot(x,y,'o',xi,yfun,xi,yipch,'--');
```

The three figures generated by the program are shown below (axes labels were added with the Plot Editor). The data points are marked with circles, the interpolation curves are plotted with dashed lines, and the function is shown with a solid line. The left figure shows the linear interpolation, the middle is the spline, and the figure on the right shows the pchip interpolation.

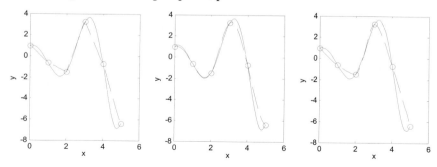

8.4 THE BASIC FITTING INTERFACE

The basic fitting interface is a tool that can be used to perform curve fitting and interpolation interactively. By using the interface the user can:

- Curve fit the data points with polynomials of various degrees up to 10, and with spline and Hermite interpolation methods.

- Plot the various fits on the plot simultaneously so that they can be compared.

- Plot the residuals of the various polynomial fits and compare the norm of the residuals.

- Calculate the value of specific points with the various fits.

- Add the equations of the polynomials to the plot.

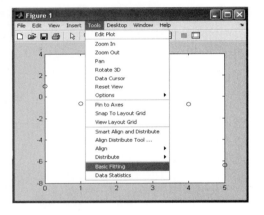

To activate the basic fitting interface, the user first has to make a plot of the data points. Then, the interface is activated by selecting **Basic Fitting** in the **Tools** menu, as shown on the right. This opens the Basic Fitting Window, shown in Figure 8-3. When the window first opens, only one panel (the **Plot fits** panel) is visible. The window can be extended to show a second panel (the **Numerical results** panel) by clicking on the → button. One click adds the first section of the panel, and a second click makes the window look as it is shown in Figure 8-3. The window can be reduced back by clicking on the ← button. The first two items in the Basic Fitting Window are related to the selection of the data points:

Select data: Used to select a specific set of data points for curve fitting in a figure that has more than one set of data points. Only one set of data points can be curve-fitted at a time, but multiple fits can be performed simultaneously on the same set.

Center and scale x data: When this box is checked, the data is centered at zero mean and scaled to unit standard deviation. This might be needed in order to improve the accuracy of numerical computation.

The next four items are in the **Plot fits** panel and are related to the display of the fit:

Check to display fits on figure: The user selects the fits to be displayed in the figure. The selection includes interpolation with spline interpolant (interpolation method) that uses the `spline` function, interpolation with Hermite interpolant that uses the `pchip` function, and polynomials of various degrees that use the

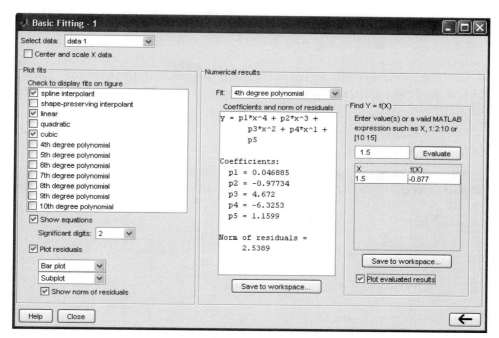

Figure 8-3: The Basic Fitting Window.

polyfit function. Several fits can be selected and displayed simultaneously.

Show equations: When this box is checked, the equations of the polynomials that were selected for the fit are displayed in the figure. The equations are displayed with the number of significant digits that is selected in the adjacent significant digits menu.

Plot residuals: When this box is checked, a plot that shows the residual at each data point is created (residuals are defined in Section 8.2.1). By selecting the proper choices in the adjacent windows, the plot can be a bar plot, a scatter plot, or a line plot, and can be displayed as a subplot in the same Figure Window that has the plot of the data points, or as a separate plot in a different Figure Window.

Show norm of residuals: When this box is checked, the norm of the residuals is displayed in the plot of the residuals. The norm of the residual is a measure of the quality of the fit. A smaller norm corresponds to a better fit.

The next three items are in the **Numerical results** panel. They provide the numerical information of one fit, independently of the fits that are displayed:

Fit: The user selects the fit he wants to examine numerically. The fit is shown on the plot only if it is selected in the **Plot fit** panel.

Coefficients and norm of residuals: Displays the numerical results for the polynomial fit that is selected in the **Fit** menu. It includes the coefficients of the polynomial and the norm of the residuals. The results can be saved by clicking on the **Save to workspace** button.

<u>**Find Y = f(x):**</u>   Provides a means for obtaining interpolated (or extrapolated) numerical values for specified values of the independent variable. Enter the value of the independent variable in the box, and click on the **Evaluate** button. When the **Plot evaluated result** box is checked, the point is displayed on the plot.

As an example, the basic fitting interface is used for fitting the data points from Sample Problem 8-3. The Basic Fitting Window is the one shown in Figure

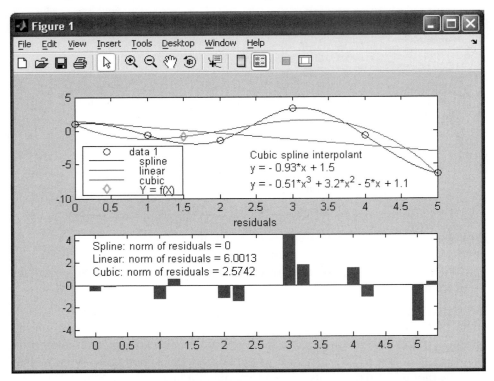

Figure 8-4: A Figure Window modified by the Basic Fitting Interface.

8-3, and the corresponding Figure Window is shown in Figure 8-4. The Figure Window includes a plot of the points, one interpolation fit (spline), two polynomial fits (linear and cubic), a display of the equations of the polynomial fits, and a mark of the point $x = 1.5$ that is entered in the **Find Y = f(x)** box of the Basic Fitting Window. The Figure Window also includes a plot of the residuals of the polynomial fits and a display of their norm.

8.5 EXAMPLES OF MATLAB APPLICATIONS

Sample Problem 8-4: Determining wall thickness of a box

The outside dimensions of a rectangular box (bottom and four sides, no top), made of aluminum, are 24 by 12 by 4 inches. The wall thickness of the bottom and the sides is x. Derive an expression that relates the weight of the box and the wall thickness x. Determine the thickness x for a box that weighs 15 lb. The specific weight of aluminum is 0.101 lb/in³.

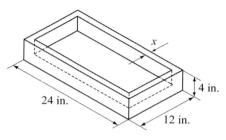

Solution

The volume of the aluminum V_{Al} is calculated from the weight W of the box by:

$$V_{Al} = \frac{W}{\gamma}$$

where γ is the specific weight. The volume of the aluminum from the dimensions of the box is given by:

$$V_{Al} = 24 \cdot 12 \cdot 4 - (24 - 2x)(12 - 2x)(4 - x)$$

where the inside volume of the box is subtracted from the outside volume. This equation can be rewritten as:

$$(24 - 2x)(12 - 2x)(4 - x) + V_{Al} - (24 \cdot 12 \cdot 4) = 0$$

which is a third degree polynomial. A root of this polynomial is the required thickness x. A program in a script file that determines the polynomial and solves for the roots is:

W = 15; gama = 0.101;	Assign W and gama.
VAlum = W/gama;	Calculate the volume of the aluminum.
a = [-2 24];	Assign the polynomial 24 – 2x to a.
b = [-2 12];	Assign the polynomial 12 – 2x to b.
c = [-1 4];	Assign the polynomial 4 – x to c.
Vin = conv(c, conv(a,b));	Multiply the three polynomials above.
polyeq = [0 0 0 (VAlum - 24*12*4)] + Vin	Add V_{Al}– 24*12*4 to Vin.
x = roots(polyeq)	Determine the roots of the polynomial.

Note in the second to last line, that in order to add the quantity $V_{Al} - (24 \cdot 12 \cdot 4)$ to the polynomial Vin it has to be written as a polynomial of the same order as Vin (Vin is a polynomial of third order). When the program (saved as

Chap8SamPro4) is executed, the coefficients of the polynomial and the value of x are displayed:

>> Chap8SamPro4

polyeq =
 -4.0000 88.0000 -576.0000 148.5149

| The polynomial is:
| $-4x^3 + 88x^2 - 576x + 148.515.$

x =
 10.8656 + 4.4831i
 10.8656 - 4.4831i
 0.2687

| The polynomial has one real root, $x = 0.2687$ in.
| which is the thickness of the aluminum wall.

Sample Problem 8-5: Floating height of a buoy

An aluminum thin-walled sphere is used as a marker buoy. The sphere has a radius of 60 cm, and a wall thickness of 12 mm. The density of aluminum is $\rho_{Al} = 2690$ kg/m³. The buoy is placed in the ocean where the density of the water is 1030 kg/m³. Determine the height h between the top of the buoy and the surface of the water.

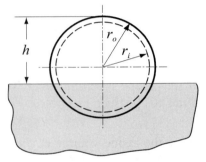

Solution

According to Archimedes law, the buoyancy force applied to an object that is placed in a fluid is equal to the weight of the fluid that is displaced by the object. Accordingly, the aluminum sphere will be at such a depth that the weight of the sphere is equal to the weight of the fluid displaced by the part of the sphere that is submerged in the water.

The weight of the sphere is given by:

$$W_{sph} = \rho_{Al} V_{Al} g = \rho_{Al} \frac{4}{3}\pi(r_o^3 - r_i^3)g$$

where V_{Al} is the volume of the aluminum, r_o and r_i are the outside and inside radii of the sphere, respectively, and g is the gravitational acceleration.

The weight of the water that is displaced by the spherical cap that is submerged in the water is given by:

$$W_{wtr} = \rho_{wtr} V_{wtr} g = \rho_{wtr} \frac{1}{3}\pi(2r_o - h)^2(r_o + h)g$$

Setting the two weights equal to each other gives the following equation:

$$h^3 - 3r_o h^2 + 4r_o^3 - 4\frac{\rho_{Al}}{\rho_{wtr}}(r_o^3 - r_i^3) = 0$$

The last equation is a third degree polynomial for h. The root of the polynomial is the answer.

A solution with MATLAB is obtained by writing the polynomials and using the roots function to determine the value of h. This is done in the script file below:

rout = 0.60; rin = 0.588;	Assign the radii to variables.
rhoalum = 2690; rhowtr = 1030;	Assign the densities to variables.
a0 = 4*rout^3 - 4*rhoalum*(rout^3 - rin^3)/rhowtr;	Assign the coefficient a0.
p = [1 -3*rout 0 a0];	Assign the coefficient vector of the polynomial.
h = roots(p)	Calculate the roots of the polynomial.

When the script file is executed in the Command Window, as shown below, the answer is three roots, since the polynomial was of the third degree. The only answer that is physically possible is the second, where $h = 0.9029$ m.

```
>> Chap8SamPro5
h =
    1.4542          The polynomial has three roots. The only one that is
    0.9029          physically possible for the problem is 0.9029 m.
   -0.5570
```

Sample Problem 8-6:   Determining the size of a capacitor

An electrical capacitor has an unknown capacitance. In order to determine its capacitance it is connected to the circuit shown. The switch is first connected to B and the capacitor is charged. Then, the switch is switched to A and the capacitor discharges through the resistor. As the capacitor is discharging, the voltage across the capacitor is measured for 10 s in intervals of 1 s. The recorded measurements are given in the table below. Plot the voltage as a function of time and determine the capacitance of the capacitor by fitting an exponential curve to the data points.

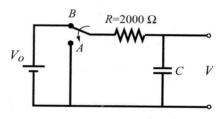

t (s)	1	2	3	4	5	6	7	8	9	10
V (V)	9.4	7.31	5.15	3.55	2.81	2.04	1.26	0.97	0.74	0.58

Solution

When a capacitor discharges through a resistor, the voltage of the capacitor as a function of time is given by:

$$V = V_0 e^{(-t)/(RC)}$$

Where V_0 is the initial voltage, R the resistance of the resistor, and C the capacitance of the capacitor. As was explained in Section 8.2.2 the exponential function can be written as a linear equation for $\ln(V)$ and t in the form:

$$\ln(V) = \frac{-1}{RC}t + \ln(V_0)$$

This equation which has the form $y = mx + b$ can be fitted to the data points by using the `polyfit(x,y,1)` function with t as the independent variable x and $\ln(V)$ as the dependent variable y. The coefficients m and b determined by the `polyfit` function are then used to determine C and V_0 by:

$$C = \frac{-t}{Rm} \quad \text{and} \quad V_0 = e^b$$

The following program written in a script file determines the best fit exponential function to the data points, determines C and V_0, and plots the points and the fitted function.

R = 2000;	Define R.
t = 1:10;	Assign the data points to vectors t and v.
v = [9.4 7.31 5.15 3.55 2.81 2.04 1.26 0.97 0.74 0.58];	
p = polyfit(t,log(v),1);	Use the `polyfit` function with t and `log(v)`.
C = -1/(R*p(1))	Calculate C from p(1), which is m in the equation.
V0 = exp(p(2))	Calculate V0 from p(2), which is b in the equation.
tplot = 0:0.1:10;	Create vector tplot of time for plotting the function.
vplot = V0*exp(-tplot./(R*C));	Create vector vplot for plotting the function.
plot(t,v,'o',tplot,vplot)	

When the script file is executed (saved as Chap8SamPro6) the values of C and V_0 are displayed in the Command Window as shown below:

```
>> Chap8SamPro6

C =
    0.0016

V0 =
    13.2796
```

The capacitance of the capacitor is 1600 µF.

The program creates also the following plot (axes labels were added to the plot using the Plot Editor):

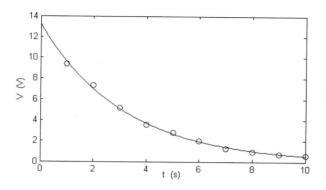

Sample Problem 8-7: Temperature dependence of viscosity

Viscosity, μ, is a property of gases and fluids that characterizes their resistance to flow. For most materials viscosity is highly sensitive to temperature. Below is a table that gives the viscosity of SAE 10W oil at different temperatures (Data from B.R. Munson, D.F. Young, and T.H. Okiishi, "Fundamental of Fluid Mechanics," 4th Edition, John Wiley and Sons, 2002). Determine an equation that can be fitted to the data.

T (°C)	−20	0	20	40	60	80	100	120
μ (N s/m²) $(\times 10^{-5})$	4	0.38	0.095	0.032	0.015	0.0078	0.0045	0.0032

Solution

To examine what type of equation might provide a good fit to the data, μ is plotted as a function of T (absolute temperature) with linear scale for T and logarithmic scale for μ. The plot, shown on the right indicates that the data points do not appear to line up along a straight line. This means that a simple exponential function of the form $y = be^{mx}$, which models a straight line in these axes, will not provide the best fit. Since the points in the figure appear to lie along a curved line, a function that can possibly have a good fit to the data is:

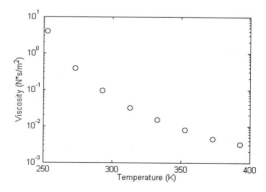

$$\ln(\mu) = a_2 T^2 + a_1 T + a_0$$

This function can be fitted to the data by using MATLAB polyfit(x,y,2) function (second degree polynomial), where the independent variable is T and the dependent variable is ln(μ). The equation above can be solved for μ to give the viscosity as a function of temperature:

$$\mu = e^{(a_2 T^2 + a_1 T + a_0)} = e^{a_0} e^{a_1 T} e^{a_2 T^2}$$

The following program determines the best fit of the function, and creates a plot that displays the data points and the function.

```
T = [-20:20:120];
mu = [4  0.38  0.095  0.032  0.015  0.0078  0.0045  0.0032];
TK = T + 273;
p = polyfit(TK,log(mu),2)
Tplot = 273 + [-20:120];
muplot = exp(p(1)*Tplot.^2 + p(2)*Tplot + p(3));
semilogy(TK,mu,'o',Tplot,muplot)
```

When the program executes (saved as Chap8SamPro7), the coefficients that are determined by the `polyfit` function are displayed in the Command Window (shown below) as three elements of the vector p.

```
>> Chap8SamPro7
p =
   0.0003  -0.2685  47.1673
```

With these coefficients the viscosity of the oil as a function of temperature is:

$$\mu = e^{(0.0003 T^2 - 0.2685 T + 47.1673)} = e^{47.1673} e^{(-0.2685) T} e^{0.0003 T^2}$$

The plot that is generated shows that the equation correlates well to the data points (axes labels were added with the Plot Editor).

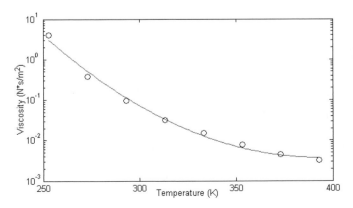

8.6 PROBLEMS

1. Plot the polynomial $y = 1.5x^4 - 5x^2 + x + 2$ in the domain $-2 \le x \le 2$. First create a vector for x, next use the `polyval` function to calculate y, and then use the `plot` function.

2. Divide the polynomial $15x^5 + 35x^4 - 37x^3 - 19x^2 + 41x - 15$ by the polynomial $5x^3 - 4x + 3$.

3. Divide the polynomial $4x^4 + 6x^3 - 2x^2 - 5x + 3$ by the polynomial $x^2 + 4x + 2$.

4. A cylindrical stainless-steel fuel tank has an outside diameter of 40 cm and a length of 70 cm. The side of the cylinder and the bottom and top ends have a thickness of x. Determine x if the mass of the tank is 18 kg. The density of stainless steel is 7920 kg/m³.

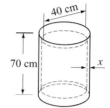

5. A gas tank is shaped as a circular cylinder with hemispheric caps at the ends as shown. The radius of the cylinder is r and its length is $4r$. Determine r if the volume of the tank is 30 ft.³.

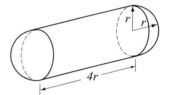

6. Write a user-defined function that adds or subtracts two polynomials of any order. Name the function `p=polyadd(p1,p2,operation)`. The first two input arguments `p1` and `p2` are the vectors of the coefficients of the two polynomials. (If the two polynomials are not of the same order the function adds the necessary zero elements to the shorter vector.) The third input argument `operation` is a string that can be either `'add'` or `'sub'`, for adding or subtracting the polynomials, respectively and the output argument is the resulting polynomial.

   Use the function to add and subtract the following polynomials:
   $f_1(x) = x^5 - 7x^4 + 11x^3 - 4x^2 - 5x - 2$ and $f_2(x) = 9x^2 - 10x + 6$.

7. Write a user-defined function that calculates the maximum (or minimum) of a quadratic equation of the form:

$$f(x) = ax^2 + bx + c$$

   Name the function `[x,y,w] = maxormin(a,b,c)`. The input arguments are the coefficients a, b, and c. The output arguments are x the coordinate of the maximum (or minimum), y the maximum (or minimum) value, and w which is equal to 1 if y is a maximum, and equal to 2 if y is a minimum.

Use the function to determine the maximum or minimum of the following functions:

a) $f(x) = 6x^2 - 18x + 6$

b) $f(x) = -4x^2 - 20x + 5$

8. A farmer wants to build three identical rectangular pens as shown in the figure. A total of 60 meters of fencing material is available. Determine the dimensions a and b that maximize the total area that is enclosed by the fence. To solve the problem the maximum of a quadratic equation must be found. Use the user-defined function from Problem 7 to find the maximum.

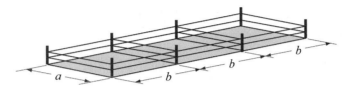

9. The following points are given:

x	−5	−4	−2.2	−1	0	1	2.2	4	5	6	7
y	0.1	0.2	0.8	2.6	3.9	5.4	3.6	2.2	3.3	6.7	8.9

a) Fit the data with a first-order polynomial. Make a plot of the points and the polynomial.

b) Fit the data with a third-order polynomial. Make a plot of the points and the polynomial.

c) Fit the data with a fourth-order polynomial. Make a plot of the points and the polynomial.

d) Fit the data with a tenth-order polynomial. Make a plot of the points and the polynomial.

10. The population of China from the year 1940 to the year 2000 is given in the following table:

Year	1940	1950	1960	1970	1980	1990	2000
Population (millions)	537	557	682	826	981	1135	1262

a) Determine the exponential function that best fits the data. Use the function to estimate the population in 1955.

b) Curve fit the data with a quadratic equation (second order polynomial). Use the function to estimate the population in 1955.

c) Fit the data with linear and spline interpolations. Estimate the population in 1955 with linear and spline interpolations.

In each part make a plot of the data points (circle markers) and the curve fitting or the interpolation curves. Note that part c has two interpolation curves. The actual population in China in 1955 was 614.4 million.

11. The standard air density, D, (average of measurements made) at different heights, h, from sea level up to a height of 33 km is given below.

h (km)	0	3	6	9	12	15
D (kg/m³)	1.2	0.91	0.66	0.47	0.31	0.19
h (km)	18	21	24	27	30	33
D (kg/m³)	0.12	0.075	0.046	0.029	0.018	0.011

a) Make the following four plots of the data points (density as a function of height). (1) both axes with linear scale, (2) h with log axis, D with linear axis, (3) h with linear axis D with log axis, (4) both with log axes. According to the plots choose a function (linear, power, exponential, or logarithmic) that can best fit the data points and determine the coefficients of the function.

b) Plot the function and the points using linear axes.

12. Write a user-defined function that fits data points to a power function of the form: $y = bx^m$. Name the function [b,m] = powerfit(x,y), where the input arguments x and y are vectors with the coordinates of the data points, and the output arguments b and m are the constants of the fitted power equation. Use the function to fit the data below. Make a plot that shows the data points and the function.

x	0.5	1.9	3.3	4.7	6.1	7.5
y	0.8	10.1	25.7	59.2	105	122

13. A thermocouple is a sensor used for measuring temperature. It is made by joining two wires made from dissimilar materials. To measure temperature, two thermocouples are connected in a circuit as shown on the right. One thermocouple is placed in a medium with known constant temperature T_{ref} (e.g. ice water), and the other is placed where the temperature T is to be measured. A voltage v is generated when the two temperatures are not the same. The voltage v can be modeled as a

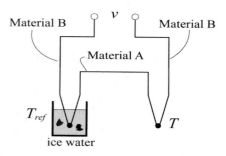

function of temperature by an expression of the form:

$$v = K_s(T - T_{ref})$$

where K_s is a constant that depends on the two materials that are used for the thermocouple.

The following are the results from an experiment that was done for determining the constant K_s of a certain thermocouple. Using this data determine K_s by using curve fitting.

T (°C)	25	100	200	300	400	500	600	700
v (mV)	1.11	4.03	8.16	12.62	16.54	20.90	23.7	29.15

14. The yield strength, σ_y, of many metals depends on the size of the grains. For these metals the relationship between the yield stress and the average grain diameter d can be modeled by the Hall-Petch equation:

$$\sigma_y = \sigma_0 + kd^{\left(\frac{-1}{2}\right)}$$

The following are results from measurements of average grain diameter and yield stress.

d (mm)	0.005	0.009	0.016	0.025	0.040	0.062	0.085	0.110
σ_y (MPa)	205	150	135	97	89	80	70	67

a) Using curve fitting, determine the constants σ_0 and k in the Hall-Petch equation for this material. Using the constants determine with the equation the yield stress of material with grain size of 0.05 mm. Make a plot that shows the data points with circle markers and the Hall-Petch equation with a solid line.

b) Use linear interpolation to determine the yield stress of material with grain size of 0.05 mm. Make a plot that shows the data points with circle markers and linear interpolation with a solid line.

c) Use cubic interpolation to determine the yield stress of material with grain size of 0.05 mm. Make a plot that shows the data points with circle markers and cubic interpolation with a solid line.

15. The ideal gas equation relates the volume, pressure, temperature, and the quantity of a gas by:

$$V = \frac{nRT}{P}$$

where V is the volume in litters, P is the pressure in atm, T is the temperature in degrees K, n is the number of moles, and R is the Gas Constant.

An experiment is conducted for determining the value of the gas constant

R. In the experiment 0.05 mol of gas is compressed to different volumes by applying pressure to the gas. At each volume the pressure and temperature of the gas is recorded. Using the data given below, determine *R* by plotting *V* versus *T/P*, and fitting the data points with a linear equation.

V (L)	0.75	0.65	0.55	0.45	0.35
T (°C)	25	37	45	56	65
P (atm)	1.63	1.96	2.37	3.00	3.96

16 Viscosity is a property of gases and fluids that characterizes their resistance to flow. For most materials viscosity is highly sensitive to temperature. For gases, the variation of viscosity with temperature is frequently modeled by the Suzerainty equation which has the form:

$$\mu = \frac{CT^{3/2}}{T + S}$$

where μ is the viscosity, *T* the absolute temperature, and *C* and *S* are empirical constants. Below is a table that gives the viscosity of air at different temperatures (Data from B.R. Munson, D.F. Young, and T.H. Okiishi, "Fundamental of Fluid Mechanics," 4th Edition, John Wiley and Sons, 2002).

T (°C)	−20	0	40	100	200	300	400	500	1000
μ (N s/m²) $(\times 10^{-5})$	1.63	1.71	1.87	2.17	2.53	2.98	3.32	3.64	5.04

Determine the constants *C* and *S* by curve fitting the equation to the data points. Make a plot of viscosity versus temperature (in °C). In the plot show the data points with markers and the curve-fitted Suzerainty equation with a solid line.

The curve fitting can be done by rewriting the equation in the form:

$$\frac{T^{3/2}}{\mu} = \frac{1}{C}T + \frac{S}{C}$$

and using a first order polynomial.

Chapter 9

Three-Dimensional Plots

Three-dimensional (3-D) plots can be a useful way to present data that consists of more than two variables. MATLAB provides various options for displaying three-dimensional data. They include line and wire, surface, mesh plots, and many others. The plots can also be formatted to have a specific appearance and special effects. Many of the three-dimensional plotting features are described in this chapter. Additional information can be found in the Help Window under **Plotting and Data Visualization**.

In many ways this chapter is a continuation of Chapter 5 where two-dimensional plots were introduced. The 3-D plots are presented in a separate chapter since not all MATLAB users use them. In addition, it was felt that for new users of MATLAB it is better to practice 2-D plotting first and learn the material in Chapters 6–8 before attempting 3-D plotting. It is assumed in the rest of this chapter that the reader is familiar with 2-D plotting.

9.1 LINE PLOTS

A three-dimensional line plot is a line that is obtained by connecting points in a three-dimensional space. A basic 3-D plot is created with the `plot3` command, which is very similar to the `plot` command, and has the form:

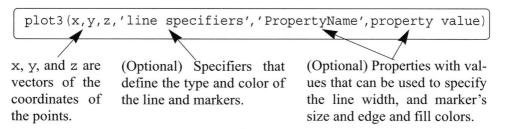

```
plot3(x,y,z,'line specifiers','PropertyName',property value)
```

x, y, and z are vectors of the coordinates of the points.

(Optional) Specifiers that define the type and color of the line and markers.

(Optional) Properties with values that can be used to specify the line width, and marker's size and edge and fill colors.

- The three vectors with the coordinates of the data points must have the same number of elements.

- The line specifiers, properties, and property values are the same as in 2-D plots (see Section 5.1).

For example, if the coordinates x, y, and z are given as a function of the parameter t by:

$$x = \sqrt{t}\sin(2t)$$
$$y = \sqrt{t}\cos(2t)$$
$$z = 0.5t$$

A plot of the points for $0 \le t \le 6\pi$ can be produced by the following script file:

```
t = 0:0.1:6*pi;
x = sqrt(t).*sin(2*t);
y = sqrt(t).*cos(2*t);
z = 0.5*t;
plot3(x,y,z,'k','linewidth',1)
grid on
xlabel('x'); ylabel('y'); zlabel('z')
```

When executed the plot that is created is shown in Figure 9-1 below:

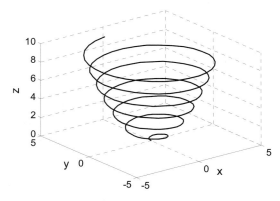

Figure 9-1: A plot of the function $x = \sqrt{t}\sin(2t)$, $y = \sqrt{t}\cos(2t)$, $z = 0.5t$ **for** $0 \le t \le 6\pi$.

9.2 MESH AND SURFACE PLOTS

Mesh and surface plots are three-dimensional plots that are used for plotting functions of the form $z = f(x, y)$ where the x and y are the independent variables and z is the dependent variable. It means that within a given domain the value of z can be calculated for any combination of x and y. Mesh and surface plots are created in three steps. The first step is to create a grid in the x-y plane that covers the domain of the function. The second step is to calculate the value of z at each point of the grid. The third step is to create the plot. The three steps are explained next.

Creating a grid in the x-y plane:

The grid is a set of points in the x-y plane in the domain of the function. The density of the grid (number of points used to define the domain) is defined by the user. Figure 9-2 shows, for example, a grid in the domain $-1 \le x \le 3$ and $1 \le y \le 4$.

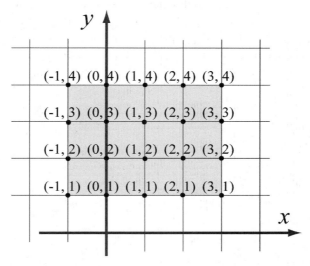

Figure 9-2: A grid in the x-y plane for the domain $-1 \le x \le 3$ and $1 \le y \le 4$ with spacing of 1.

In this grid the distance between the grid points is one unit. The points of the grid can be defined by two matrices X and Y. Matrix X has the x coordinates of all the points, and matrix Y has the y coordinates of all the points:

$$X = \begin{bmatrix} -1 & 0 & 1 & 2 & 3 \\ -1 & 0 & 1 & 2 & 3 \\ -1 & 0 & 1 & 2 & 3 \\ -1 & 0 & 1 & 2 & 3 \end{bmatrix} \quad \text{and} \quad Y = \begin{bmatrix} 4 & 4 & 4 & 4 & 4 \\ 3 & 3 & 3 & 3 & 3 \\ 2 & 2 & 2 & 2 & 2 \\ 1 & 1 & 1 & 1 & 1 \end{bmatrix}$$

The X matrix is made from identical rows since in each row of the grid the points have the same x coordinate. In the same way the Y matrix is made from identical columns since in each column of the grid the y coordinate of the points is the same.

MATLAB has a built-in function, called meshgrid, that can be used for creating the *X* and *Y* matrices. The form of the `meshgrid` function is:

$$[X,Y] = \text{meshgrid}(x,y)$$

X is the matrix of the *x* coordinates of the grid points.
Y is the matrix of the *y* coordinates of the grid points.

x is a vector that divides the domain of *x*.
y is a vector that divides the domain of *y*.

In the vectors x and y the first and the last elements are the respective boundaries of the domain. The density of the grid is determined by the number of elements in the vectors. For example, the mesh matrices *X* and *Y* that correspond to the grid in Figure 9-2 can be created with the `meshgrid` command by:

```
>> x = -1:3;
>> y = 1:4;
>> [X,Y] = meshgrid(x,y)
X =
    -1   0   1   2   3
    -1   0   1   2   3
    -1   0   1   2   3
    -1   0   1   2   3
Y =
     1   1   1   1   1
     2   2   2   2   2
     3   3   3   3   3
     4   4   4   4   4
```

Once the grid matrices exist, they can be used for calculating the value of *z* at each grid point.

Calculating the value of *z* at each point of the grid:

The value of *z* at each point is calculated by using element-by-element calculations in the same way it is used with vectors. When the independent variables *x* and *y* are matrices (must be of the same size) the calculated dependent variable is also a matrix of the same size. The value of *z* at each address is calculated from the corresponding values of *x* and *y*. For example, if *z* is given by:

$$z = \frac{xy^2}{x^2 + y^2}$$

the value of *z* at each point of the grid above is calculated by:

```
>> Z = X.*Y.^2./(X.^2 + Y.^2)
```

```
Z =
   -0.5000        0   0.5000   0.4000   0.3000
   -0.8000        0   0.8000   1.0000   0.9231
   -0.9000        0   0.9000   1.3846   1.5000
   -0.9412        0   0.9412   1.6000   1.9200
```

Once the three matrices exist, they can be used to plot mesh or surface plots.

Making mesh and surface plots:

A mesh or a surface plot is created with the `mesh` or `surf` commands, which have the forms:

mesh(X,Y,Z)	surf(X,Y,Z)

where X, Y, are matrices with the coordinates of the grid and Z is a matrix with the value of z at the grid points. The mesh plot is made of lines that connect the points. In the surface plot, areas within the mesh lines are colored.

As an example, the following script file contains a complete program that creates the grid and then makes a mesh (or surface) plot of the function $z = \dfrac{xy^2}{x^2 + y^2}$ over the domain $-1 \le x \le 3$ and $1 \le y \le 4$.

```
x = -1:0.1:3;
y = 1:0.1:4;
[X,Y] = meshgrid(x,y);
Z = X.*Y.^2./(X.^2 + Y.^2);
mesh(X,Y,Z)                    Type surf(X,Y,Z) for surface plot.
xlabel('x'); ylabel('y'); zlabel('z')
```

Note that in the program above the vectors x and y have a much smaller spacing than the spacing earlier in the section. The smaller spacing creates a denser grid. The figures created by the program are:

Additional comments on the mesh command:

- The plots that are created have colors that vary according to the magnitude of z. The variation in color adds to the three-dimensional visualization of the plots. The color can be changed to be a constant either by using the Plot Editor in the Figure Window (select the edit arrow, click on the figure to open the Property Editor Window, then change the color in the Mesh Properties list), or by using the `colormap(C)` command. In this command C is a three-element vector in which the first, second, and third elements specify the intensity of Red, Green, and Blue (RGB) colors, respectively. Each element can be a number between 0 (minimum intensity) and 1 (maximum intensity). Some typical colors are:

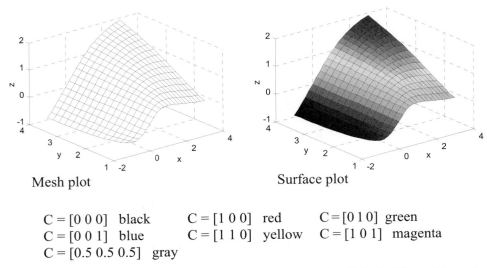

Mesh plot   Surface plot

C = [0 0 0]   black   C = [1 0 0]   red   C = [0 1 0]   green
C = [0 0 1]   blue   C = [1 1 0]   yellow   C = [1 0 1]   magenta
C = [0.5 0.5 0.5]   gray

- When the mesh command executes, the grid is on by default. The grid can be turned off with the `grid off` command.

- A box can be drawn around the plot with the `box on` command.

- The `mesh` and `surf` commands can also be used with the form `mesh(Z)` and `surf(z)`. In this case the values of Z are plotted as a function of their addresses in the matrix. The row number is on the x axis and the column number is on the y axis.

There are several additional plotting commands that are similar to the `mesh` and `surf` commands that create plots with different features. Table 9-1 shows a summary of mesh and surface plotting commands. All the examples in the table are plots of the function $z = 1.8^{-1.5\sqrt{x^2+y^2}}\sin(x)\cos(0.5y)$ over the domain $-3 \le x \le 3$ and $-3 \le y \le 3$.

Table 9-1: Mesh and surface plots

Plot type	Example of plot	Program

Mesh Plot
 	
Function format:
`mesh(X,Y,Z)`
		`x = -3:0.25:3;`
		`y = -3:0.25:3;`
		`[X,Y] = meshgrid(x,y);`
		`Z = 1.8.^(-1.5*sqrt(X.^2 + Y.^2)).*cos(0.5*Y).*sin(X);`
		`mesh(X,Y,Z)`
		`xlabel('x'); ylabel('y')`
		`zlabel('z')`

Table 9-1: Mesh and surface plots (Continued)

Plot type	Example of plot	Program

Surface Plot
 	
Function format:
`surf(X,Y,Z)`
		x = -3:0.25:3;
		y = -3:0.25:3;
		[X,Y] = meshgrid(x,y);
		Z = 1.8.^(-1.5*sqrt(X.^2 + Y.^2)).*cos(0.5*Y).*sin(X);
		surf(X,Y,Z)
		xlabel('x'); ylabel('y')
		zlabel('z')

Mesh Curtain Plot (draws a curtain around the mesh)
 	
Function format:
`meshz(X,Y,Z)`
		x = -3:0.25:3;
		y = -3:0.25:3;
		[X,Y] = meshgrid(x,y);
		Z = 1.8.^(-1.5*sqrt(X.^2 + Y.^2)).*cos(0.5*Y).*sin(X);
		meshz(X,Y,Z)
		xlabel('x'); ylabel('y')
		zlabel('z')

Mesh and Contour Plot (draws a contour plot beneath the mesh)
 	
Function format:
`meshc(X,Y,Z)`
		x = -3:0.25:3;
		y = -3:0.25:3;
		[X,Y] = meshgrid(x,y);
		Z = 1.8.^(-1.5*sqrt(X.^2 + Y.^2)).*cos(0.5*Y).*sin(X);
		meshc(X,Y,Z)
		xlabel('x'); ylabel('y')
		zlabel('z')

Surface and Contour Plot (draws a contour plot beneath the surface)
 	
Function format:
`surfc(X,Y,Z)`
		x = -3:0.25:3;
		y = -3:0.25:3;
		[X,Y] = meshgrid(x,y);
		Z = 1.8.^(-1.5*sqrt(X.^2 + Y.^2)).*cos(0.5*Y).*sin(X);
		surfc(X,Y,Z)
		xlabel('x'); ylabel('y')
		zlabel('z')

Table 9-1: Mesh and surface plots (Continued)

Plot type	Example of plot	Program

Surface Plot with Lighting
 	
Function format:
`surfl(X,Y,Z)`
		x = -3:0.25:3;
		y = -3:0.25:3;
		[X,Y] = meshgrid(x,y);
		Z = 1.8.^(-1.5*sqrt(X.^2 + Y.^2)).*cos(0.5*Y).*sin(X);
		surfl(X,Y,Z)
		xlabel('x'); ylabel('y')
		zlabel('z')

Waterfall Plot (draws a mesh in one direction only)
 	
Function format:
`water-fall(X,Y,Z)`
		x = -3:0.25:3;
		y = -3:0.25:3;
		[X,Y] = meshgrid(x,y);
		Z = 1.8.^(-1.5*sqrt(X.^2 + Y.^2)).*cos(0.5*Y).*sin(X);
		waterfall(X,Y,Z)
		xlabel('x'); ylabel('y')
		zlabel('z')

3-D Contour Plot
 	
Function format:
`contour3(X, Y,Z,n)`
 	
n is the number of contour levels (optional)
		x = -3:0.25:3;
		y = -3:0.25:3;
		[X,Y] = meshgrid(x,y);
		Z = 1.8.^(-1.5*sqrt(X.^2 + Y.^2)).*cos(0.5*Y).*sin(X);
		contour3(X,Y,Z,15)
		xlabel('x'); ylabel('y')
		zlabel('z')

2-D Contour Plot (draws projections of contour levels on the *x-y* plane)
Function format:
`contour (X,Y,Z,n)`
n is the number of contour levels (optional)
		x = -3:0.25:3;
		y = -3:0.25:3;
		[X,Y] = meshgrid(x,y);
		Z = 1.8.^(-1.5*sqrt(X.^2 + Y.^2)).*cos(0.5*Y).*sin(X);
		contour(X,Y,Z,15)
		xlabel('x'); ylabel('y')
		zlabel('z')

9.3 PLOTS WITH SPECIAL GRAPHICS

MATLAB has additional functions for creating various types of special three-dimensional plots. A complete list can be found in the Help Window under Plotting and Data Visualization. Several of these 3-D plots are presented in Table 9-2.

Table 9-2: Specialized 3-D plots

Plot type	Example of plot	Program

Plot a Sphere
 	
Function format:
`sphere`
Returns the x, y, and z coordinates of a unit sphere with 20 faces.
`sphere(n)`
Same as above with n faces.
		sphere
		
		or:
		
		[X,Y,Z] = sphere(20);
		surf(X,Y,Z)

Plot a Cylinder
 	
Function format:
`[X,Y,Z] =`
`cylinder(r)`
Returns the x, y, and z coordinates of cylinder with profile r.
		t = linspace(0,pi,20);
		r = 1 + sin(t);
		[X,Y,Z] = cylinder(r);
		surf(X,Y,Z)
		axis square

3-D Bar Plot Function format: `bar3(Y)` Each element in Y is one bar. Columns are grouped together.		Y = [1 6.5 7; 2 6 7; 3 5.5 7; 4 5 7; 3 4 7; 2 3 7; 1 2 7]; bar3(Y)

Table 9-2: Specialized 3-D plots (Continued)

Plot type	Example of plot	Program

3-D Stem Plot (draws sequential points with markers and vertical lines from the *x-y* plane)
 	
Function format:
stem3 (X,Y,Z)
		t = 0:0.2:10;
		x = t;
		y = sin(t);
		z = t.^1.5;
		stem3(x,y,z,'fill')
		grid on
		xlabel('x'); ylabel('y')
		zlabel('z')

3-D Scatter Plot
 	
Function format:
scatter3 (X, Y, Z)
		t = 0:0.4:10;
		x = t;
		y = sin(t);
		z = t.^1.5;
		scatter3(x,y,z,'filled')
		grid on
		colormap([0.1 0.1 0.1])
		xlabel('x'); ylabel('y')
		zlabel('z')

3-D Pie Plot Function format: pie3 (X, explode)		X = [5 9 14 20]; explode = [0 0 1 0]; pie3(X,explode) explode is a vector (same length as X) of 0's and 1's. 1 offsets the slice from the center.

The examples in the table do not show all the options available with each plot type. More details for each type of plot can be obtained in the Help Window, or by typing help command_name in the Command Window.

9.4 *THE* view *COMMAND*

The view command controls the direction from which the plot is viewed. This is done by specifying a direction in terms of azimuth and elevation angles, as seen in Figure 9-3, or by defining a point in space from which the plot is viewed. To set the viewing angle of the plot, the `view` command has the form:

> `view(az,el) or view([az,el])`

- `az` is the azimuth, which is an angle (in degrees) in the *x*-*y* plane measured relative to the negative *y* axis direction and defined as positive in the counterclockwise direction.

- `el` is the angle of elevation (in degrees) from the *x*-*y* plane. A positive value corresponds to opening an angle in the direction of the *z* axis.

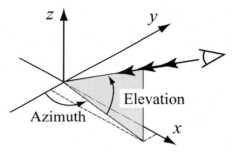

Figure 9-3: Azimuth and Elevation angles.

- The default view angles are $az = -37.5^o$, and $el = 30^o$.

For example, the surface plot from Table 9-1 is plotted again in Figure 9-4 below with a viewing angles of $az = 20^o$, and $el = 35^o$.

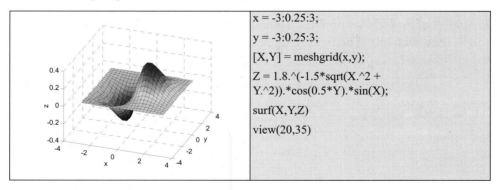

Figure 9-4: A surface plot of the function $z = 1.8^{-1.5\sqrt{x^2+y^2}}\sin(x)\cos(0.5y)$ **with viewing angles of** $az = 20^o$, **and** $el = 35^o$.

- By choosing appropriate azimuth and elevation angles the `view` command can be used to plot projections of 3-D plots on various planes according to the following table:

Projection plane	**_az_ value**	**_el_ value**
_x-y_ (top view)	0	90
_x-z_ (side view)	0	0
_y-z_ (side view)	90	0

An example of a top view is shown next. The figure shows the top view of the function that is plotted in Figure 9-1.

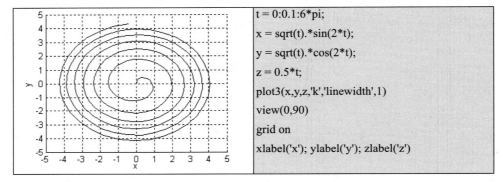

```
t = 0:0.1:6*pi;
x = sqrt(t).*sin(2*t);
y = sqrt(t).*cos(2*t);
z = 0.5*t;
plot3(x,y,z,'k','linewidth',1)
view(0,90)
grid on
xlabel('x'); ylabel('y'); zlabel('z')
```

Figure 9-5: A top view plot of the function $x = \sqrt{t}\sin(2t)$, $y = \sqrt{t}\cos(2t)$, $z = 0.5t$ **for** $0 \le t \le 6\pi$.

Examples of projections onto the _x-z_ and _y-z_ planes are shown next in Figures 9-6 and 9-7, respectively. The figures show mesh plot projections of the function plotted in Table 9-1.

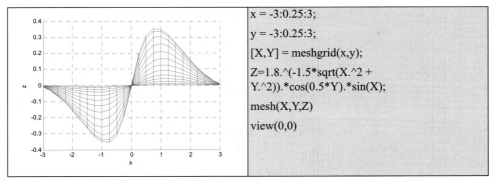

```
x = -3:0.25:3;
y = -3:0.25:3;
[X,Y] = meshgrid(x,y);
Z=1.8.^(-1.5*sqrt(X.^2 +
Y.^2)).*cos(0.5*Y).*sin(X);
mesh(X,Y,Z)
view(0,0)
```

Figure 9-6: Projections on the _x-z_ plane of the function $z = 1.8^{-1.5\sqrt{x^2+y^2}}\sin(x)\cos(0.5y)$.

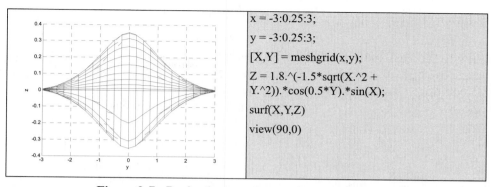

	x = -3:0.25:3;
	y = -3:0.25:3;
	[X,Y] = meshgrid(x,y);
	Z = 1.8.^(-1.5*sqrt(X.^2 + Y.^2)).*cos(0.5*Y).*sin(X);
	surf(X,Y,Z)
	view(90,0)

Figure 9-7: Projections on the *y-z* planes of the function

$$z = 1.8^{-1.5\sqrt{x^2+y^2}}\sin(x)\cos(0.5y).$$

- The view command can also set a default view:

  view(2)   sets the default to the top view which is a projection onto the *x-y* plane with $az = 0°$, and $el = 90°$.

  view(3)   sets the default to the standard 3-D view with $az = -37.5°$, and $el = 30°$.

- The viewing direction can also be set by selecting a point in space from which the plot is viewed. In this case the view command has the form view([x,y,z]), where x, y, and z are the coordinates of the point. The direction is determined by the direction from the specified point to the origin of the coordinate system and is independent of the distance. This means that the view is the same with point [6, 6, 6] as with point [10, 10, 10]. Top view can be set up with [0, 0, 1]. A side view of the *x-z* plane from the negative *y* direction can be set with [0, –1, 0], and so on.

9.5 EXAMPLES OF MATLAB APPLICATIONS

Sample Problem 9-1: 3-D projectile trajectory

A projectile is fired with an initial velocity of 250 m/s at an angle of $\theta = 65°$ relative to the ground. The projectile is aimed directly North. Because of a strong wind blowing to the West, the projectile also moves in this direction at a constant speed of 30 m/s. Determine and plot the trajectory of the projectile until it hits the ground. For comparison, plot also (in the same figure) the trajectory that the projectile would have had if there was no wind.

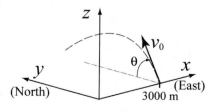

Solution

As shown in the figure, the coordinate system is set up such that the x and y axes point to the East and North directions, respectively. Then, the motion of the projectile can be analyzed by considering the vertical direction z and the two horizontal components x and y. Since the projectile is fired directly North, the initial velocity v_0 can be resolved into a horizontal y component and a vertical z component:

$$v_{0y} = v_0\cos(\theta) \quad \text{and} \quad v_{0z} = v_0\sin(\theta)$$

In addition, due to the wind the projectile has a constant velocity in the negative x direction, $v_x = -30$ m/s.

The initial position of the projectile (x_0, y_0, z_0) is at point $(3000, 0, 0)$. In the vertical direction the velocity and position of the projectile are given by:

$$v_z = v_{0z} - gt \quad \text{and} \quad z = z_0 + v_{0z}t - \frac{1}{2}gt^2$$

The time it takes the projectile to reach the highest point $(v_z = 0)$ is $t_{hmax} = \frac{v_{0z}}{g}$.

The total flying time is twice this time, $t_{tot} = 2t_{hmax}$. In the horizontal direction the velocity is constant (both in the x and y directions), and the position of the projectile is given by:

$$x = x_0 + v_x t \quad \text{and} \quad y = y_0 + v_{0y}t$$

The following MATLAB program written in a script file solves the problem by following the equations above.

```
v0 = 250; g = 9.81; theta = 65;
x0 = 3000; vx = -30;
v0z = v0*sin(theta*pi/180);
v0y = v0*cos(theta*pi/180);
t = 2*v0z/g;
tplot = linspace(0,t,100);                  Creating a time vector with 100 elements.
z = v0z*tplot - 0.5*g*tplot.^2;
y = v0y*tplot;                               Calculating the x, y and z coordinates
                                             of the projectile at each time.
x = x0 + vx*tplot;
xnowind(1:length(y)) = x0;                   Constant x coordinate when no wind.
plot3(x,y,z,'k-',xnowind,y,z,'k--')          Two 3-D line plots.
grid on
axis([0 6000 0 6000 0 2500])
xlabel('x (m)'); ylabel('y (m)'); zlabel('z (m)')
```

The figure generated by the program is shown below.

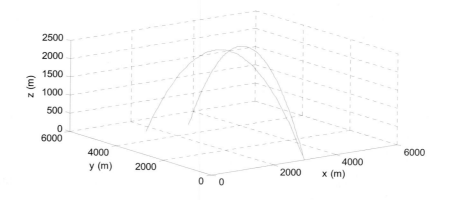

Sample Problem 9-2: Electric potential of two point charges

The electric potential V around a charged particle is given by:

$$V = \frac{1}{4\pi\varepsilon_0}\frac{q}{r}$$

where $\varepsilon_0 = 8.8541878 \times 10^{-12}\dfrac{C}{N \cdot m^2}$ is the permittivity constant, q is the magnitude of the charge in Coulombs, and r is the distance from the particle in meters. The electric field of two or more particles is calculated by using superposition. For example, the electric potential at a point due to two particles is given by:

$$V = \frac{1}{4\pi\varepsilon_0}\left(\frac{q_1}{r_1} + \frac{q_2}{r_2}\right)$$

where q_1, q_2, r_1, and r_2 are the charges of the particles and the distance from the point to the corresponding particle, respectively.

Two particles with a charge of $q_1 = 2 \times 10^{-10}$ C and $q_2 = 3 \times 10^{-10}$ C are positioned in the x-y plane at points (0.25, 0, 0) and (−0.25, 0, 0), respectively as shown. Calculate and plot the electric potential due to the two particles at points in the x-y plane that are located in the domain $-0.2 \le x \le 0.2$ and $-0.2 \le y \le 0.2$ (the units in the x-y plane are in meters). Make the plot such that the x-y plane is the plane of the points, and the z axis is the magnitude of the electric potential.

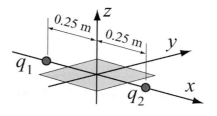

Solution

The problem is solved following these steps:

a) A grid is created in the *x-y* plane with the domain $-0.2 \le x \le 0.2$ and $-0.2 \le y \le 0.2$.

b) The distance from each grid point to each of the charges is calculated.

c) The electric potential at each point is calculated.

d) The electric potential is plotted.

The following is a program in a script file that solves the problem.

```
eps0 = 8.85e-12; q1 = 2e-10; q2 = 3e-10;
k = 1/(4*pi*eps0);
x = -0.2:0.01:0.2;
y = -0.2:0.01:0.2;
[X,Y] = meshgrid(x,y);            Creating a grid in the x-y plane.
r1 = sqrt((X+0.25).^2 + Y.^2);    Calculating the distance r₁ for each grid point.
r2 = sqrt((X-0.25).^2 + Y.^2);    Calculating the distance r₂ for each grid point.
V = k*(q1./r1 + q2./r2);     Calculating the electric potential V at each grid point.
mesh(X,Y,V)
xlabel('x (m)'); ylabel('y (m)'); zlabel('V (V)')
```

The plot generated when the program runs is:

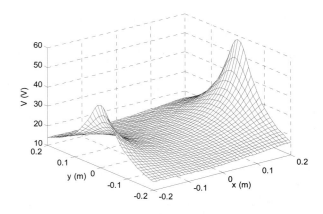

Sample Problem 9-3: Heat conduction in a square plate

Three sides of a rectangular plate ($a = 5$ m, $b = 4$ m) are kept at a temperature of $0\,^\circ$C and one side is kept at a temperature of $T_1 = 80\,^\circ$C, as shown in the figure. Determine and plot the temperature distribution $T(x,y)$ in the plate.

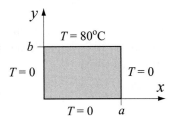

Solution

The temperature distribution, $T(x,y)$ in the plate can be determined by solving the two-dimensional heat equation. For the given boundary conditions $T(x,y)$ can be expressed analytically by Fourier series (Erwin Kreyszig, "Advanced Engineering Mathematics," John Wiley and Sons, 1993):

$$T(x, y) = \frac{4T_1}{\pi} \sum_{n=1}^{\infty} \frac{\sin\left[(2n-1)\frac{\pi x}{a}\right] \sinh\left[(2n-1)\frac{\pi y}{a}\right]}{(2n-1) \quad \sinh\left[(2n-1)\frac{\pi b}{a}\right]}$$

A program in a script file that solves the problem is listed below. The program follows these steps:

a) Create an X, Y grid in the domain $0 \le x \le a$ and $0 \le y \le b$. The length of the plate, a, is divided into 20 segments, and the width of the plate, b, is divided into 16 segments.

b) Calculate the temperature at each point of the mesh. The calculations are done point by point using a double loop. At each point the temperature is determined by adding k terms of the Fourier series.

c) Make a surface plot of T.

```
a = 5; b = 4; na = 20; nb = 16; k = 5; T0 = 80;
clear T
x = linspace(0,a,na);
y = linspace(0,b,nb);
[X,Y] = meshgrid(x,y);              Creating a grid in the x-y plane.
for i = 1:nb                        First loop, i is the index of the grid's row.
  for j = 1:na                      Second loop, j is the index of the grid's column.
    T(i,j) = 0;
    for n = 1:k                     Third loop, n is the nth term of the Fourier
      ns = 2*n - 1;                 series, k is the number of terms.
    T(i,j) = T(i,j) + sin(ns*pi*X(i,j)/a).*sinh(ns*pi*Y(i,j)/a)/(sinh(ns*pi*b/a)*ns);
    end
    T(i,j) = T(i,j)*4*T0/pi;
  end
end
mesh(X,Y,T)
xlabel('x (m)'); ylabel('y (m)'); zlabel('T ( ^oC)')
```

The program was executed twice. First using 5 terms ($k = 5$) in the Fourier series to calculate the temperature at each point, and then with $k = 50$. The mesh plots created in each execution are shown in the figures below. The temperature should

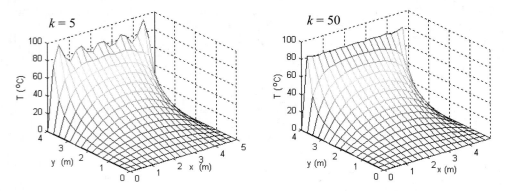

be uniformly 80°C at $y = 4$ m. Note the effect of the number of terms (k) on the accuracy at $y = 4$ m.

9.6 PROBLEMS

1. The position of a moving particle as a function of time is given by:

$$x = (2 + 4\cos(t))\cos(t)$$
$$y = (2 + 4\cos(t))\sin(t)$$
$$z = t^2$$

    Plot the position of the particle for $0 \leq t \leq 20$.

2. A circular staircase can be modeled by the parametric equations:

$$x = R\cos\left(2\pi n\frac{t}{h}\right)$$
$$y = R\sin\left(2\pi n\frac{t}{h}\right)$$
$$z = \frac{t}{h}$$

    where R is the radius of the staircase, h is the height of the floors, and n is the number of revolutions that the staircase makes in each floor. A building has 2 floors with $h = 3$ m. Make plots of two possible staircases. One with $R = 1.5$ m and $n = 3$, and the other with $R = 4$ m and $n = 2$. Plot the two staircases in the same figure.

3. The ladder of a firetruck can be ele-
   vated (increase of angle ϕ), rotated
   about the z axis (increase of angle θ),
   and extended (increase of r). Initially
   the ladder rests on the truck, $\phi = 0$,
   $\theta = 0$, and $r = 8$ m. Then the ladder is
   moved to a new position by raising the
   ladder at a rate of 5 deg/s, rotating at a
   rate of 8 deg/s, and extending the ladder

   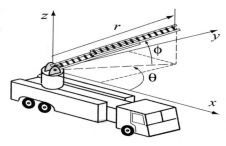

   at a rate of 0.6 m/s. Determine and plot the position of the tip of the ladder for
   10 seconds.

4. Make a 3-D surface plot and a 3-D contour plot of the function $z = -\dfrac{x^2}{4} - \dfrac{y^2}{4}$
   in the domain $-4 \le x \le 4$ and $-4 \le y \le 4$.

5. Make a 3-D surface plot and a contour plot (both in the same figure) of the
   function $z = (y + 3)^2 + 1.5x^2 - x^2y$ in the domain $-3 \le x \le 3$ and $-3 \le y \le 3$.

6. The ideal gas law relates the pressure, temperature, and volume of a gas by:

   $$P = \frac{nRT}{V}$$

   where P is the pressure in Pa, n is the number of moles, $R = 8.31$ J/mol-K is
   the gas constant, T is the temperature in degrees K, and V is the volume in m³.
   Make a 3-D plot that shows the variation of pressure (dependent variable,
   z axis) with volume (independent variable, x axis) and temperature (independent variable, y axis) of one mole of gas. The domains for the volume and
   temperature are: $0.5 \times 10^{-3} \le V \le 2 \times 10^{-3}$ m³, and $273 \le T \le 473$ K.

7. Molecules of a gas in a container are moving around at different speeds. Max-
   well's speed distribution law gives the probability distribution $P(v)$ as a function of temperature and speed:

   $$P(v) = 4\pi\left(\frac{M}{2\pi RT}\right)^{3/2} v^2 e^{(-Mv^2)/(2RT)}$$

   where M is the molar mass of the gas in kg/mol, $R = 8.31$ J/mol-K is the gas
   constant, T is the temperature in degrees K, and v is the molecules speed in
   m/s.

   Make a 3-D plot of $P(v)$ as a function of v and T for $0 \le v \le 1000$ m/s and
   $70 \le T \le 320$ K for oxygen gas (molar mass 0.032 kg/mol).

8. An *RLC* circuit with an alternating voltage source is shown. The source voltage v_s is given by $v_s = v_m \sin(\omega_d t)$ where $\omega_d = 2\pi f_d$ in which f_d is the driving frequency. The amplitude of the current, I, in this circuit is given by:

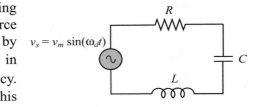

$$I = \frac{v_m}{\sqrt{R^2 + (\omega_d L - 1/(\omega_d C))^2}}$$

where R and C are the resistance of the resistor and capacitance of the capacitor, respectively. For the circuit in the figure $C = 15 \times 10^{-6}$ F, $L = 240 \times 10^{-3}$ H, and $v_m = 24$ V.

a) Make a 3-D plot of I (*z* axis) as a function of ω_d (*x* axis) for $60 \leq f \leq 110$ Hz, and a function of R (*y* axis) for $10 \leq R \leq 40 \ \Omega$.

b) Make a plot that is a projection on the *x-z* plane. Estimate from this plot the natural frequency of the circuit (the frequency at which I is maximum). Compare the estimate with the calculated value of $1/(2\pi\sqrt{LC})$.

9. The normal stress, σ_{xx}, due to bending moments M_z and M_y at a point (y, z) in the cross section of a rectangular beam is given by:

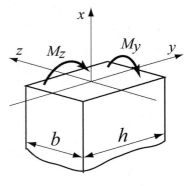

$$\sigma_{xx} = \frac{-M_z y}{I_{zz}} + \frac{M_y z}{I_{yy}}$$

where I_{zz} and I_{yy} are the area moments of inertia defined by:

$$I_{zz} = \frac{1}{12}bh^3 \text{ and } I_{yy} = \frac{1}{12}hb^3 .$$

Determine and plot the normal stress in the cross-sectional area shown in the figure, given that: $h = 40$ mm, $b = 30$ mm, $M_y = 2500$ N-m, and $M_z = 3600$ N-m. Plot the coordinates y and z in the horizontal plane, and the normal stress in the vertical direction.

10. A defect in a crystal lattice where a raw of atoms is missing is called an edge dislocation. The stress field around an edge dislocation is given by:

$$\sigma_{xx} = \frac{-Gb}{2\pi(1-v)}\frac{y(3x^2+y^2)}{(x^2+y^2)^2}$$

$$\sigma_{yy} = \frac{Gb}{2\pi(1-v)}\frac{y(x^2-y^2)}{(x^2+y^2)^2}$$

$$\tau_{xy} = \frac{Gb}{2\pi(1-v)}\frac{x(x^2-y^2)}{(x^2+y^2)^2}$$

where G is the shear modulus, b is the Burgers vector, and v is Poisson's ratio. Plot the stress components (each in a separate figure) due to an edge dislocation in aluminum for which $G = 27.7 \times 10^9$ Pa, $b = 0.286 \times 10^{-9}$ m, and $v = 0.334$. Plot the stresses in the domain $-5 \times 10^{-9} \le x \le 5 \times 10^{-9}$ m and $-5 \times 10^{-9} \le y \le -1 \times 10^{-9}$ m. Plot the coordinates x and y in the horizontal plane, and the stresses in the vertical direction.

Chapter 10
Applications in Numerical Analysis

Numerical methods are commonly used for solving mathematical problems that are formulated in science and engineering where it is difficult or even impossible to obtain exact solutions. MATLAB has a large library of functions for numerically solving a wide variety of mathematical problems. This chapter explains how to use a number of the most frequently used of these functions. It should be pointed out here that the purpose of this book is to show users how to use MATLAB. Some general information on the numerical methods is given, but the details, which can be found in books on numerical analysis, are not included.

The following topics are presented in this chapter: solving an equation with one unknown, finding a minimum or a maximum of a function, numerical integration, and solving a first-order ordinary differential equation.

10.1 SOLVING AN EQUATION WITH ONE VARIABLE

An equation with one variable can be written in the form $f(x) = 0$. The solution is the value of x where the function crosses the x axis (the value of the function is zero), which means that the function changes sign at x. An exact solution is a value of x for which the value of the function is exactly zero. If such a value does not exist or is difficult to determine, a numerical solution can be determined by finding an x that is very close to the point where the function changes its sign (crosses the x-axis). This is done by the iterative process where in each iteration the computer determines a value of x that is closer to the solution. The iterations stop when the difference in x between two iterations is smaller than some measure. In general, a function can have none, one, several, or infinite number of solutions.

In MATLAB a zero of a function can be determined with the command (built-in function) `fzero` that has the form:

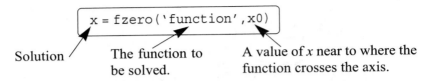

Solution   The function to   A value of x near to where the
                be solved.   function crosses the axis.

Additional details on the arguments of `fzero`:

- x is the solution, which is a scalar.

- `'function'` is the function to be solved. It can be entered in three different ways:
  1. The simplest way is to enter the mathematical expression as a string.
  2. The function can also be created as a user-defined function in a function file (refer to Chapter 6), then the name of the function is typed as a string.
  3. The function can be created as an inline function (see Section 6.8) and its name then typed as a string.

- The function has to be written in a standard form. For example, if the function to be solved is $xe^{-x} = 0.2$, it has to be written as $f(x) = xe^{-x} - 0.2 = 0$. If this function is entered into the `fzero` command as a string, it is typed as: `'x*exp(-x)-0.2'`.

- When a function is entered as a string, it cannot include predefined variables. For example, if the function to be entered is $f(x) = xe^{-x} - 0.2$, it is not possible to define b=0.2 and then enter `'x*exp(-x)-b'`.

- x0 can be a scalar or a two-element vector. If it is entered as a scalar, it has to be a value of x near the point where the function crosses the x axis. If x0 is entered as a vector, the two elements have to be points on opposite sides of the solution such that $f(x0(1))$ has a different sign than $f(x0(2))$. When a function has more than one solution, each solution can be determined separately by using the `fzero` function and entering values for x0 that are near each of the solutions.

- A good way to find approximately where a function has a solution is to make a plot of the function. In many applications in science and engineering the domain of the solution can be estimated. Often when a function has more than one solution only one of the solutions will have a physical meaning.

Sample Problem 10-1: Solving a nonlinear equation

Determine the solution of the equation $xe^{-x} = 0.2$.

Solution

The equation is first written in a form of a function: $f(x) = xe^{-x} - 0.2$. A plot of the function, shown on the right, shows that the function has one solution between 0 and 1 and another solution between 2 and 3. The plot is obtained by typing

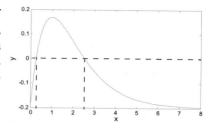

```
>> fplot('x*exp(-x) - 0.2',[0 8])
```

in the Command Window. The solutions of the function are found by using the fzero command twice; once with a value of x0 between 0 and 1, (e.g. x0 = 0.7), and once with x0 between 2 and 3, (e.g, x0 = 2.8). This is shown below:

```
>> x1 = fzero('x*exp(-x) - 0.2',0.7)
x1 =
    0.2592
```
The first solution is 0.2592.

```
>> x1 = fzero('x*exp(-x) - 0.2',2.8)
x1 =
    2.5426
```
The second solution is 2.5426.

Additional comments:

- The fzero command finds zeros of a function only where the function crosses the x-axis. The command does not find a zero at points where the function touches but does not cross the x-axis.

- If a solution cannot be determined, NaN is assigned to x.

- The fzero command has additional options (see the Help Window). Two of the more important options are:
  [x fval]=fzero('function', x0) assigns the value of the function at x to the variable fval.
  x=fzero('function',x0,optimset('display','iter')) displays the output of each iteration during the process of finding the solution.

- When the function can be written in the form of a polynomial, the solution, or the roots, can be found with the roots command, as is explained in Chapter 8 (Section 8.1.2).

- The fzero command can also be used to find the value of x where the function has a specific value. This is done by translating the function up or down. For example, in the function of Sample Problem 10-1 the first value of x where the function is equal to 0.1 can be determined by solving the equation $xe^{-x} - 0.3 = 0$. This is shown below:

```
>> x = fzero('x*exp(-x) - 0.3',0.5)
x =
    0.4894
```

10.2 FINDING A MINIMUM OR A MAXIMUM OF A FUNCTION

In many applications there is a need to determine the local minimum or the maximum of a function of the form $y = f(x)$. In calculus the value of x that corresponds to a local minimum or the maximum is determined by finding the zero of the derivative of the function. The value of y is determined by substituting the x into the function. In MATLAB the value of x where a one-variable function $f(x)$ within the interval $x_1 \le x \le x_2$ has a minimum can be determined with the fminbnd command which has the form:

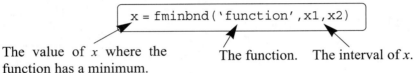

```
x = fminbnd('function',x1,x2)
```

The value of x where the function has a minimum.   The function.   The interval of x.

- The function can be entered as a string, as the name of a function file, or as the name of an inline function, in the same way as with the fzero command. See Section 10.1 for details.

- The value of the function at the minimum can be added to the output by using the option:

```
[x fval]=fminbnd('function',x1,x2)
```

where the value of the function at x is assigned to the variable fval.

- Within a given interval, the minimum of a function can either be at one of the end points of the interval, or at a point within the interval where the slope of the function is zero (local minimum). When the fminbnd command is executed, MATLAB looks for a local minimum. If a local minimum is found, its value is compared to the value of the function at the end points of the interval. MATLAB returns the point with the actual minimum value of the interval.

For example, let's consider the function $f(x) = x^3 - 12x^2 + 40.25x - 36.5$ which is plotted in the interval $0 \le x \le 8$ in the figure on the right. It can be observed that there is a local minimum between 5 and 6, and that the absolute minimum is at $x = 0$. Using the fminbnd command with the interval $3 \le x \le 8$ to find the

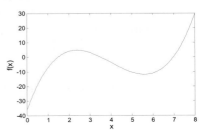

location of the local minimum and the value of the function at this point gives:

```
>> [x fval] = fminbnd('x^3 - 12*x^2 + 40.25*x - 36.5',3,8)
x =
    5.6073
fval =
   -11.8043
```

> The local minimum is at $x = 5.6073$. The value of the function at this point is -11.8043.

Notice that the `fminbnd` command gives the local minimum. If the interval is changed to $0 \le x \le 8$, the `fminbnd` gives:

```
>> [x fval] = fminbnd('x^3 - 12*x^2 + 40.25*x - 36.5',0,8)
x =
    0
fval =
   -36.5000
```

> The minimum is at $x = 0$. The value of the function at this point is -36.5.

For this interval the `fminbnd` command gives the absolute minimum which is at the end point $x = 0$.

- The `fminbnd` command can also be used to find the maximum of a function. This is done by multiplying the function by -1 and finding the minimum. For example, the maximum of the function $f(x) = xe^{-x} - 0.2$, (from Sample Problem 10-1) in the interval $0 \le x \le 8$ can be determined by finding the minimum of the function $f(x) = -xe^{-x} + 0.2$ as shown below:

```
>> [x fval] = fminbnd('-x*exp(-x) + 0.2',0,8)
x =
    1.0000
fval =
   -0.1679
```

> The maximum is at $x = 1.0$. The value of the function at this point is 0.1679.

10.3 NUMERICAL INTEGRATION

Integration is a common mathematical operation in science and engineering. Calculating area and volume, velocity from acceleration, work from force and displacement are just a few examples where integrals are used. Integration of simple functions can be done analytically, but more involved functions are frequently difficult or impossible to integrate analytically. In calculus courses the integrand (the quantity to be integrated) is usually a function. In applications of science and engineering the integrand can be a function or a set of data points. For example, data points from discrete measurements of flow velocity can be used to calculate volume.

It is assumed in the presentation below that the reader has knowledge of integrals and integration. A definite integral of a function $f(x)$ from a to b has the form:

$$q = \int_a^b f(x)\,dx$$

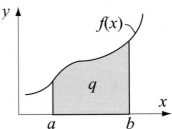

The function $f(x)$ is called the integrand and the numbers a and b are the limits of integration. Graphically, the value of the integral q is the area between the graph of the function, the x axis, and the limits a and b (the shaded area in the figure). When a definite integral is calculated analytically $f(x)$ is always a function. When the integral is calculated numerically $f(x)$ can be a function or a set of points. In numerical integration the total area is obtained by dividing the area into small sections, calculating the area of each section, and adding them up. Various numerical methods have been developed for this purpose. The difference between the methods is in the way that the area is divided into sections and the method by which the area of each section is calculated. The reader is referred to books on numerical analysis for details of the numerical techniques.

The following describes how to use the three MATLAB built-in integration functions quad, quadl and trapz. The quad and quadl commands are used for integration when $f(x)$ is a function, while the trapz is used when $f(x)$ is given by data points.

The quad command:

The form of the quad command which uses the adaptive Simpson method of integration is:

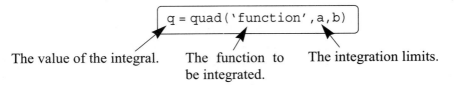

The value of the integral.   The function to   The integration limits.
                                be integrated.

- The function can be entered as a string, as the name of a function file, or the name of an inline function, in the same way as with the fzero command. See Section 10.1 for details. The first two methods are demonstrated in Sample Problem 10-2 on the following page.

- The function $f(x)$ must be written for an argument x that is a vector (use element-by-element operations), such that it calculates the value of the function for each element of x.

- The user has to make sure that the function does not have a vertical asymptote between a and b.

- quad calculates the integral with an absolute error that is smaller than 1.0e–6. This number can be changed by adding an optional tol argument to the command:

  q = quad('function',a,b,tol)

  tol is a number that defines the maximum error. With larger tol the integral is calculated less accurately, but faster.

The quadl command:

The form of the quadl (the last letter is a lower case L) command is exactly the same as the quad command:

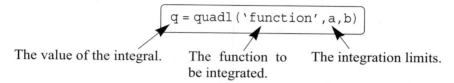

q = quadl('function',a,b)

The value of the integral. The function to be integrated. The integration limits.

All the comments that are listed above for the quad command are valid for the quadl command. The difference between the two commands is in the numerical method that is used for calculating the integration. The quadl uses the adaptive Lobatto method, which can be more efficient for high accuracies and smooth integrals.

Sample Problem 10-2: Numerical integration of a function

Use numerical integration to calculate the following integral:

$$\int_0^8 (xe^{-x^{0.8}} + 0.2)dx$$

Solution

For illustration, a plot of the function for the interval $0 \le x \le 8$ is shown on the right. The solution uses the quad command and shows how to enter the function in the command in two ways. In the first, it is entered directly by typing the expression as an argument. In the second, a function file is created first and its name is subsequently entered in the command.

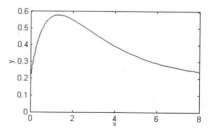

The use of the quad command in the Command Window with the function to be integrated typed in as a string, is shown below. Note that the function is typed with element-by-element operations.

```
>> quad('x.*exp(-x.^0.8) + 0.2',0,8)
```

```
ans =
   3.1604
```

The second method is to first create a function file that calculates the function to be integrated. The function file (named `y=Chap10Sam2(x)`) is:

```
function y = Chap10Sam2(x)
y = x.*exp(-x.^0.8) + 0.2;
```

Note again that the function is written with element-by-element operations such that the argument x can be a vector. The integration is then done in the Command Window by typing the command `quad` with the name of the function file as the argument `'function'` as shown below:

```
>> q = quad('Chap10Sam2',0,8)
q =
   3.1604
```

The `trapz` command:

The `trapz` command can be used for integrating a function that is given as data points. It uses the numerical trapezoidal method of integration. The form of the command is:

$$\boxed{\texttt{q = trapz(x,y)}}$$

where x and y are vectors with the *x* and *y* coordinates of the points, respectively. The two vectors must be of the same length.

10.4 ORDINARY DIFFERENTIAL EQUATIONS

Differential equations play a crucial role in science and engineering since they are in the foundation of virtually every physical phenomena that is involved in engineering applications. Only a limited number of differential equations can be solved analytically. Numerical methods, on the other hand, can give an approximate solution to almost any equation. Obtaining a numerical solution might not be, however, a simple task. This is because a numerical method that can solve any equation does not exist. Instead, there are many methods that are suitable for solving different types of equations. MATLAB has a large library of tools that can be used for solving differential equations. To fully utilize the power of MATLAB, however, requires that the user have knowledge of differential equations and the various numerical methods that can be used for solving them.

      This section describes in detail how to use MATLAB to solve a first order ordinary differential equation. The possible numerical methods that can be used for solving such an equation are mentioned and described in general terms, but are not explained from a mathematical point of view. This section provides information for solving simple, "nonproblematic," first-order equations. This solution provides the basis for solving higher-order equations and a system of equations.

An ordinary differential equation (ODE) is an equation that contains an independent variable, a dependent variable, and derivatives of the dependent variable. The equations that are considered here are of first order with the form:

$$\frac{dy}{dx} = f(x, y)$$

where x and y are the independent and dependent variables, respectively. A solution is a function $y = f(x)$ that satisfies the equation. In general, many functions can satisfy a given ODE and more information is required for determining the solution of a specific problem. The additional information is the value of the function (the dependent variable) at some value of the independent variable.

Steps for solving a single first order ODE:

For the remainder of this section the independent variable is taken as t (time). This is done because in many applications time is the independent variable, and to be consistent with the information in the **Help** menu of MATLAB.

Step 1: **Write the problem in a standard form.**

Write the equation in the form:

$$\frac{dy}{dt} = f(t, y) \quad \text{for } t_0 \le t \le t_f, \text{ with } \quad y = y_0 \text{ at } t = t_0.$$

As shown above, three pieces of information are needed for solving a first order ODE: An equation that gives an expression for the derivative of y with respect to t, the interval of the independent variable, and the initial value of y. The solution is the value of y as a function of t between t_0 and t_f.

An example of a problem to solve is:

$$\frac{dy}{dt} = \frac{t^3 - 2y}{t} \quad \text{for } 1 \le t \le 3 \text{ with } y = 4.2 \text{ at } t = 1.$$

Step 2: **Create a function file.**

Create a user-defined function (function file) that calculates $\frac{dy}{dt}$ for given values of t and y. For the example problem above, the function file is:

```
function dydt = ODEexp1(t,y)
dydt = (t^3 - 2*y)/t;
```

Step 3: **Select a method of solution.**

Select the numerical method that you would like MATLAB to use in the solution. Many numerical methods have been developed to solve first order ODE's, and several of the methods are available as built-in functions in MATLAB. In a typical numerical method, the time interval is divided into small time steps. The solution starts at the known point y_0, and then by using one of the integration methods the value of y is calculated at each time step. Table 10-1 lists seven ODE solver commands, which are MATLAB built-in functions that can be used for solving a first order ODE. A short description of each solver is included in the table.

Table 10-1: MATLAB ODE Solvers

ODE Solver Name	Description
ode45	For nonstiff problems, one step solver, best to apply as a first try for most problems. Based on explicit Runge-Kutta method.
ode23	For nonstiff problems, one step solver. Based on explicit Runge-Kutta method. Often quicker but less accurate than ode45.
ode113	For nonstiff problems, multistep solver.
ode15s	For stiff problems, multistep solver. Use if ode45 failed. Uses a variable order method.
ode23s	For stiff problems, one step solver. Can solve some problems that ode15 cannot.
ode23t	For moderately stiff problems.
ode23tb	For stiff problems. Often more efficient than ode15s.

In general, the solvers can be divided into two groups according to their ability to solve stiff problems and according to whether they use one or multistep methods. Stiff problems are ones that include fast and slowly changing components and require small time steps in their solution. One-step solvers use information from one point to obtain solution at the next point. Multistep solvers use information from several previous points to find the solution at the next point. The details of the different methods are beyond the scope of this book.

It is impossible to know ahead of time which solver is the most appropriate for a specific problem. A suggestion is to first try ode45 which gives good results for many problems. If a solution is not obtained because the problem is stiff, it is suggested to try the solver ode15s.

Step 3: **Solve the ODE.**

The form of the command that is used to solve an initial value ODE problem is the same for all the solvers and all the equations that are solved. The form is:

```
[t,y] = solver_name('ODEfun',tspan,y0)
```

Additional information:

`solver_name`	Is the name of the solver (numerical method) that is used (e.g. `ode45` or `ode23s`)
`'ODEfun'`	Is the name, entered as a string, of the user-defined function (function file) that calculates $\frac{dy}{dt}$ for given values of t and y.
`tspan`	A vector that specifies the interval of the solution. The vector must have at least two elements, but can have more. If the vector has only two elements, the elements must be `[t0 tf]`, which are the initial and final points of the solution interval. The vector `tspan` can have, however, additional points between the first and last points. The number of elements in `tspan` affects the output from the command. See `[t,y]` below.
`y0`	Is the initial value of y (the value of y at the first point of the interval.)
`[t,y]`	Is the output, which is the solution of the ODE. `t` and `y` are column vectors. The first and the last points are the beginning and end points of the interval. The spacing and number of points in between depends on the input vector `tspan`. If `tspan` has two elements (the beginning and end points) the vectors `t` and `y` contain the solution at every integration step calculated by the solver. If `tspan` has more than two points (additional points between the first and the last) the vectors `t` and `y` contain the solution only at these points. The number of points in `tspan` does not affect the time steps used for the solution by the program.

For example, a solution to the problem stated in Step 1:

$$\frac{dy}{dt} = \frac{t^3 - 2y}{t} \quad \text{for } 1 \le t \le 3 \quad \text{with} \quad y = 4.2 \text{ at } t = 1,$$

can be obtained by:

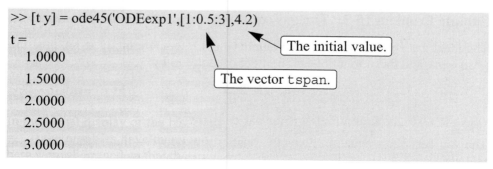

```
>> [t y] = ode45('ODEexp1',[1:0.5:3],4.2)
t =
   1.0000
   1.5000
   2.0000
   2.5000
   3.0000
```

The initial value.

The vector `tspan`.

y =
  4.2000
  2.4528
  2.6000
  3.7650
  5.8444

The solution is obtained with the solver `ode45`. The name of the user-defined function from step 2 is `'ODEexp1'`. The solution starts at $t = 1$ and ends at $t = 3$ with increments of 0.5 (according to the vector `tspan`). To show the solution, the problem is solved again below using `tspan` with smaller spacing, and the solution is plotted with the `plot` command.

```
>> [t y] = ode45('ODEexp1',[1:0.01:3],4.2);
>> plot(t,y)
>> xlabel('t'), ylabel('y')
```

The plot that is created is:

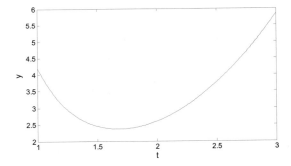

10.5 EXAMPLES OF MATLAB APPLICATIONS

Sample Problem 10-3: The gas equation

The ideal gas equation relates the volume (V in L), temperature (T in K), pressure (P in atm), and the amount of gas (number of moles n) by:

$$p = \frac{nRT}{V}$$

where $R = 0.08206$ L-atm/mol-K is the gas constant.

The van der Waals equation gives the relationship between these quantities for a real gas by:

$$\left(P + \frac{n^2 a}{V^2}\right)(V - nb) = nRT$$

where a and b are constants that are specific for each gas.

Use the `fzero` function to calculate the volume of 2 mol CO_2 at temperature of 50°C, and pressure of 6 atm. For CO_2, $a = 3.59$ L^2-atm/mol^2, and $b = 0.0427$ L/mol.

Solution

The solution written in a script file is shown below.

```
global P T n a b R
R = 0.08206;
P = 6; T = 323.2; n = 2; a = 3.59; b = 0.047;
Vest = n*R*T/P;           Calculating an estimated value for V.
V = fzero('Waals',Vest)
```

The program first calculates an estimated value of the volume using the ideal gas equation. This value is then used in the `fzero` command for the estimate of the solution. The van der Waals equation is written as a user-defined function named `Waals`, which is shown below:

```
function fofx = Waals(x)
global P T n a b R
fofx = (P + n^2*a/x^2)*(x - n*b) - n*R*T;
```

In order for the script and function files to work correctly, the variables P, T, n, a, b, and R are declared global. When the script file (saved as Chap10SamPro3) is executed in the Command Window the value of V is displayed, as shown next:

```
>> Chap10SamPro3
V =
    8.6613            The volume of the gas is 8.6613 L.
```

Sample Problem 10-4: Maximum viewing angle

To get the best view of a movie, a person has to sit at a distance x from the screen such that the viewing angle θ is maximum. Determine the distance x for which θ is maximum for the configuration shown in the figure.

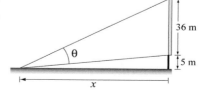

Solution

The problem is solved by writing a function for the angle θ in terms of x, and then finding the x for which the angle is maximum. In the triangle that includes θ, one side is given (the height of the screen), and the other two sides can be written in terms of x, as shown in the

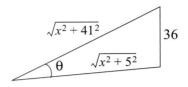

figure. One way in which θ can be written in terms of x is by using the law of cosines.

$$\cos(\theta) = \frac{(x^2 + 5^2) + (x^2 + 41^2) - 36^2}{2\sqrt{x^2 + 5^2}\sqrt{x^2 + 41^2}}$$

The angle θ is expected to be between 0 and $\pi/2$. Since $\cos(0) = 1$ and the cosine is decreasing with increasing θ, the maximum angle corresponds to the smallest $\cos(\theta)$. A plot of $\cos(\theta)$ as a function of x shows that the function has a minimum between 10 and 20. The commands for the plot are:

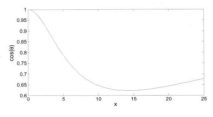

```
>>fplot('((x^2 + 5^2) + (x^2 + 41^2) - 36^2)/(2*sqrt(x^2 + 5^2)*sqrt(x^2 +
                                                41^2))',[0 25])

>> xlabel('x'); ylabel('cos(\theta)')
```

The minimum can be determined with the `fminbnd` command:

```
>> [x anglecos] = fminbnd( '((x^2 + 5^2) + (x^2 + 41^2) - 36^2)/
                            (2*sqrt(x^2 + 5^2)*sqrt(x^2 + 41^2))',10,20)

x =
   14.3178
anglecos =
   0.6225
```

> The minimum is at $x = 14.3178$ m. At this point $\cos(\theta) = 0.6225$.

```
>> angle = anglecos*180/pi
angle =
   35.6674
```

> In degrees the angle is $35.6674°$.

Sample Problem 10-5: Water flow in a river

To estimate the amount of water that flows in a river during a year, a section of the river is made up to have a rectangular cross section as shown. In the beginning of every month (starting at January 1st) the height h of the water, and the speed v of the water flow is measured. The first day of measurement is taken as 1, and the last day which is January 1st of the next year is day 366. The following data was measured:

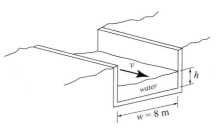

Day	1	32	60	91	121	152	182	213	244	274	305	335	366
h (m)	2.0	2.1	2.3	2.4	3.0	2.9	2.7	2.6	2.5	2.3	2.2	2.1	2.0
v (m/s)	2.0	2.2	2.5	2.7	5	4.7	4.1	3.8	3.7	2.8	2.5	2.3	2.0

Use the data to calculate the flow rate, and then integrate the flow rate to obtain an estimate of the total amount of water that flows in the river during a year.

Solution

The flow rate, Q, (volume of water per second) at each data point is obtained by multiplying the water speed by the width and height of the cross-sectional area of the water that flows in the channel:

$$Q = vwh \quad (m^3/s)$$

The total amount of water that flows is estimated by the integral:

$$V = (60 \cdot 60 \cdot 24) \int_{t_1}^{t_2} Q \, dt$$

The flow rate is given in cubic meters per second which means that the time must have units of seconds. Since the data is given in terms of days, the integral is multiplied by $(60 \cdot 60 \cdot 24)$ s/day.

The following is a program written in a script file that first calculates Q and then carries out the integration using the `trapz` command. The program also generates a plot of the flow rate vs. time.

```
w = 8;
d = [1 32 60 91 121 152 182 213 244 274 305 335 366];
h = [2 2.1 2.3 2.4 3.0 2.9 2.7 2.6 2.5 2.3 2.2 2.1 2.0];
speed = [2 2.2 2.5 2.7 5 4.7 4.1 3.8 3.7 2.8 2.5 2.3 2];
Q = speed.*w.*h;
Vol = 60*60*24*trapz(d,Q);
```

```
fprintf('The estimated amount of water that flows in the river in a year is %g
cubic meters.',Vol)

plot(d,Q)

xlabel('Day'), ylabel('Flow Rate (m^3/s)')
```

When the file (saved as Chap10SamPro5) is executed in the Command Window, the estimated amount of water is displayed and the plot is generated. Both are shown below:

>> Chap10SamPro5

The estimated amount of water that flows in the river in a year is 2.03095e+009 cubic meters.

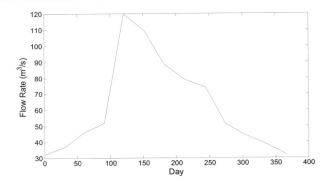

Sample Problem 10-6: Car crash into a safety bumper

A safety bumper is placed at the end of a racetrack to stop out-of-control cars. The bumper is designed such that the force that the bumper applies to the car is a function of the velocity v and the displacement x of the front edge of the bumper according to the equation:

$$F = Kv^3(x+1)^3$$

where $K = 30$ s-kg/m^5 is a constant.

A car with a mass m of 1500 kg hits the bumper at a speed of 90 km/h. Determine and plot the velocity of the car as a function of its position for $0 \leq x \leq 3$ m.

Solution

The deceleration of the car once it hits the bumper can be calculated from Newton's second law of motion.

$$ma = -Kv^3(x+1)^3$$

which can be solved for the acceleration a as a function of v and x:

$$a = \frac{-Kv^3(x+1)^3}{m}$$

The velocity as a function of x can be calculated by substituting the acceleration in the equation:

$$vdv = adx$$

which gives:

$$\frac{dv}{dx} = \frac{-Kv^2(x+1)^3}{m}$$

The last equation is a first order ODE that needs to be solved for the interval $0 \leq x \leq 3$ with the initial condition: $v = 90$ km/h at $x = 0$.

A numerical solution of the differential equation with MATLAB is shown in the following program which is written in a script file:

```
global k m
k = 30; m = 1500; v0 = 90;
xspan = [0:0.2:3];                    A vector that specifies the interval of the solution.
v0mps = v0*1000/3600;                 Changing the units of v0 to m/s.
[x v] = ode45('bumper',xspan,v0mps)   Solving the ODE.
plot(x,v)
xlabel('x (m)'); ylabel('velocity (m/s)')
```

The function file with the differential equation, named bumper, is listed below:

```
function dvdx = bumper(x,v)
global k m
dvdx = -(k*v^2*(x + 1)^3)/m;
```

When the script file executes (was saved as Chap10SamPro6) the vectors x and v are displayed in the Command Window (actually they are displayed on the screen one after the other, but to save room they are displayed below next to each other).

```
>> Chap10SamPro6

x =               v =
       0              25.0000
  0.2000              22.0420
  0.4000              18.4478
  0.6000              14.7561
  0.8000              11.4302
```

1.0000	8.6954
1.2000	6.5733
1.4000	4.9793
1.6000	3.7960
1.8000	2.9220
2.0000	2.2737
2.2000	1.7886
2.4000	1.4226
2.6000	1.1435
2.8000	0.9283
3.0000	0.7607

The plot of the velocity as a function of distance generated by the program is:

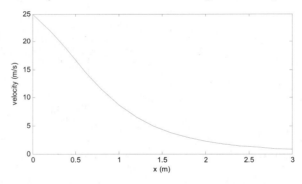

10.6 PROBLEMS

1. Determine the solution of the equation: $\cos(x) = 2x^3$

2. Determine the first three positive roots of the equation: $2\sin(x) - \sqrt{x} = -2.5$.

3. Determine the first three positive roots of the equation:
$4\cos(2x) - e^{0.5x} + 5 = 0$.

4. A box of mass $m = 20$ kg is being pulled by a rope. The force that is required to move the box is given by:
$$F = \frac{\mu mg}{\cos\theta + \mu\sin\theta}$$

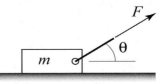

where $\mu = 0.45$ is the friction coefficient and $g = 9.81$ m/s². Determine the angle θ, if the pulling force is 92 N.

5. A scale is made of two springs, as shown in the figure. Initially, the springs are not stretched. When an object is attached to the ring, the springs stretch and the ring is displaced down a distance x. The weight of the object can be expressed in terms of the distance x by:

$$W = \frac{2K}{L}(L - L_0)(b + x)$$

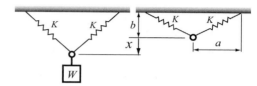

where $L_0 = \sqrt{a^2 + b^2}$ is the initial length of a spring, and

$L = \sqrt{a^2 + (b + x)^2}$ is the stretched length of the spring.

For the given scale $a = 0.15$ m, $b = 0.05$ m, and the springs' constant is $K = 2800$ N/m. Determine the distance x when a 250 N object is attached to the scale.

6. Consider again the box that is being pulled in Problem 4. Determine the angle θ at which the force that is required to pull the box is the smallest. What is the magnitude of this force?

7. A cylindrical paper cup is designed to hold 200 mL. Determine the radius and height of the cup such that the least amount of paper will be used for making the cup.

8. Determine the dimensions (radius r and height h) of the cylinder with the largest volume that can be made out of a sphere with a radius R of 15 cm.

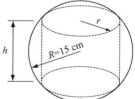

9. Planck's radiation law gives the spectral radiancy R as a function of the wave length λ and temperature T (in degrees K):

$$R = \frac{2\pi c^2 h}{\lambda^5} \frac{1}{e^{(hc)/(\lambda kT)} - 1}$$

where $c = 3.0 \times 10^8$ m/s is the speed of light, $h = 6.63 \times 10^{-34}$ J-s is the Planck constant, and $k = 1.38 \times 10^{-23}$ J/K is the Boltzmann constant.

Plot R as a function of λ for $0.2 \times 10^{-6} \le \lambda \le 6.0 \times 10^{-6}$ m at $T = 1500$ K, and determine the wavelength that gives the maximum R at this temperature.

10. Use MATLAB to calculate the following integral:

$$\int_0^5 \frac{1}{0.8x^2 + 0.5x + 2} \, dx$$

11. Use MATLAB to calculate the following integral:

$$\int_0^\pi \cos^2(0.5x)\sin^4(0.5x)\,dx$$

12. The electric field E due to a charged circular disk at a point at a distance z along the axis of the disk is given by:

$$E = \frac{\sigma z}{4\varepsilon_0} \int_0^R (z^2 + r^2)^{-3/2}(2r)\,dr$$

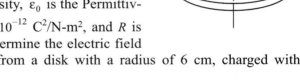

where σ is the charge density, ε_0 is the Permittivity constant, $\varepsilon_0 = 8.85 \times 10^{-12}$ C^2/N-m^2, and R is the radius of the disk. Determine the electric field at a point located 5 cm from a disk with a radius of 6 cm, charged with $\sigma = 300\ \mu$C/m^2.

13. The variation of gravitational acceleration g with altitude y is given by:

$$g = \frac{R^2}{(R+y)^2} g_0$$

where $R = 6371$ km is the radius of the earth, and $g_0 = 9.81$ m/s^2 is the gravitational acceleration at sea level. The change in the gravitational potential energy, ΔU of an object that is raised up from the Earth is given by:

$$\Delta U = \int_0^h mg\,dy$$

Determine the change in the potential energy of a satellite with a mass of 500 kg that is raised from the surface of the earth to a height of 800 km.

14. The surface area of a lake is estimated by measuring the width of the lake at intervals of 100 m. The measurements are shown in the figure. Use numerical integration to estimate the area of the lake.

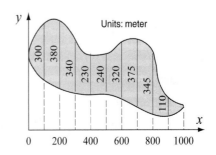

15. Solve:

$$\frac{dy}{dx} = \frac{x^2 - 2x + 3}{y^2} \quad \text{for } 0.5 \le x \le 3 \quad \text{with } y(0.5) = 2.$$

Plot the solution.

16. Solve:

$$\frac{dy}{dx} = 0.2xy + 0.5y^2 \quad \text{for } 0 \le x \le 4 \quad \text{with } y(0) = -0.5.$$

Plot the solution.

17. A water tank shaped as an inverted frustum cone has a circular hole at the bottom
on the side, as shown. According to Torri-
celli's law, the speed v of the water that is
discharging from the hole is given by:

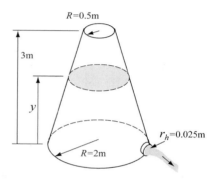

$$v = \sqrt{2gh}$$

where h is the height of the water and
$g = 9.81$ m/s². The rate at which the
height, y, of the water in the tank changes
as the water flows out through the hole is
given by:

$$\frac{dy}{dt} = \frac{\sqrt{2gy}\,r_h^2}{(2 - 0.5y)^2}$$

where r_h is the radius of the hole.
   Solve the differential equation for y. The initial height of the water is
$y = 2$ m. Solve the problem for different times and find the time where
$y = 0.1$ m. Make a plot of y as a function of time.

18. An airplane uses a parachute and
other means of braking as it slows
down on the runway after land-
ing. Its acceleration is given by
$a = -0.0035v^2 - 3$ m/s².   Since
$a = \dfrac{dv}{dt}$, the rate of change of the
velocity is given by:

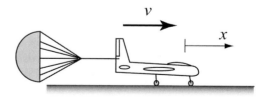

$$\frac{dv}{dt} = -0.0035v^2 - 3$$

Consider an airplane with a velocity of 300 km/h that opens its parachute and
starts decelerating at $t = 0$ s.

a) By solving the differential equation, determine and plot the velocity as a function of time from $t = 0$ s until the airplane stops.

b) Use numerical integration to determine the distance x the airplane travels as a function of time. Make a plot of x vs. time.

19. An RC circuit includes a voltage source v_s, a resistor $R = 50$ Ω, and a capacitor $C = 0.001$ F, as shown in the figure. The differential equation that describes the response of the circuit is:

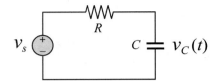

$$\frac{dv_c}{dt} + \frac{1}{RC}v_c = \frac{1}{RC}v_s$$

where v_c is the voltage of the capacitor. Initially, $v_s = 0$, and then at $t = 0$ the voltage source is changed. Determine the response of the circuit for the following three cases:

a) $v_s = 12$ V for $t \geq 0$.

b) $v_s = 12\sin(2 \cdot 60\pi t)$ V for $t \geq 0$.

c) $v_s = 12$ V for $0 \leq t \leq 0.01$ s, and then $v_s = 0$ for $t \geq 0.01$ s (rectangular pulse).

Each case corresponds to a different differential equation. The solution is the voltage of the capacitor as a function of time. Solve each case for $0 \leq t \leq 0.2$ s. For each case plot v_s and v_c vs. time (make two separate plots on the same page).

Chapter 11
Symbolic Math

All the mathematical operations done with MATLAB in the first ten chapters were numerical. The operations were carried out by writing numerical expressions that could contain numbers and variables with preassigned numerical values. When a numerical expression is executed by MATLAB, the outcome is also numerical (a single number or an array with numbers). The number, or numbers, are either exact or a floating point-approximated value. For example, typing 1/4 gives 0.2500 – an exact value, and typing 1/3 gives 0.3333 – an approximated value.

Many applications in math, science, and engineering require symbolic operations, which are mathematical operations with expressions that contain symbolic variables (variables that don't have specific numerical values when the operation is executed). The result of such operations is also a mathematical expression in terms of the symbolic variables. One simple example is solving an algebraic equation which contains several variables for one variable in terms of the others. If a, b, and x are symbolic variables, and $ax - b = 0$, x can be solved in terms of a and b to give $x = b / a$. Other examples of symbolic operations are analytical differentiation or integration of mathematical expressions. For instance, the derivative of $2t^3 + 5t - 8$ with respect to t is $6t^2 + 5$.

MATLAB has the capability of carrying out many types of symbolic operations. The numerical part of the symbolic operation is carried out by MATLAB exactly without approximating numerical values. For example, the result of adding $\frac{x}{4}$ and $\frac{x}{3}$ is $\frac{7}{12}x$ and not $0.5833x$.

Symbolic operations can be performed by MATLAB when the Symbolic Math Toolbox is installed. The Symbolic Math Toolbox is a collection of MATLAB functions that are used for execution of symbolic operations. The commands and functions for the symbolic operations have the same style and syntax as those for the numerical operations. The symbolic operations themselves are executed primarily by Maple®, which is mathematical software designed for this purpose. The Maple software is embedded within MATLAB and is automatically activated when a symbolic MATLAB function is executed. Maple also exists as separate independent software. That software, however, has a completely different structure and commands than MATLAB. The Symbolic Math Toolbox is included in the student version of MATLAB. In the standard version, the toolbox is purchased separately. To check if the Symbolic Math Toolbox is installed on a computer, the user can type the command `ver` in the Command Window. In response, MAT-LAB displays information about the version that is used as well as a list of the tool boxes that are installed.

The starting point for symbolic operations is symbolic objects. Symbolic objects are made of variables and numbers that, when used in mathematical expressions, tell MATLAB to execute the expression symbolically. Typically, the user first defines (creates) the symbolic variables (objects) that are needed, and then uses them to create symbolic expressions which are subsequently used in symbolic operation. If needed, symbolic expressions can be used in numerical operations

The first section in this chapter describes how to define symbolic objects and how to use them to create symbolic expressions. The second section shows how to change the form of existing expressions. Once a symbolic expression exists, it can be used in mathematical operations. MATLAB has a large selection of functions for this purpose. The next four sections (11.3–11.6) describe how to use MATLAB to solve algebraic equations, to carry out differentiation and integration, and to solve differential equations. Section 11.7 covers plotting symbolic expressions. How to use symbolic expressions in subsequent numerical calculations is explained in the following section.

11.1 SYMBOLIC OBJECTS, AND SYMBOLIC EXPRESSIONS

A symbolic object can be a variable (without a preassigned numerical value), a number, or an expression made of symbolic variables and numbers. Symbolic expression is a mathematical expression containing one or more symbolic objects. When typed, a symbolic expression may look like a standard numerical expression. However, because the expression contains symbolic objects, it is executed by MATLAB symbolically.

11.1.1 Creating Symbolic Objects

Symbolic objects can be variables or numbers. They can be created with the `sym` and/or the `syms` commands. A single symbolic object can be created with the `sym` command:

```
object_name = sym('string')
```

where the string which is the symbolic object is assigned to a name. The string can be:

• A single letter, or a combination of several letters (no spaces). Examples: `'a'`, `'x'`, or `'yad'`.

- A combination of letters and digits starting with a letter and with no spaces Examples: `'xh12'`, or `'r2d2'`.

- A number. Examples: `'15'`, or `'4'`.

In the first two cases (when the string is a single letter, a combination of several letters, or a combination of letters and digits), the symbolic object is a symbolic variable. In this case it is convenient (but not necessary) to name the object the same as the string. For example, *a*, *bb*, and *x*, can be defined as symbolic variables by:

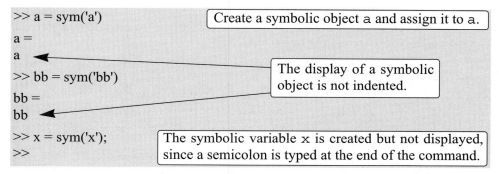

```
>> a = sym('a')          Create a symbolic object a and assign it to a.

a =

a                        The display of a symbolic
>> bb = sym('bb')        object is not indented.

bb =

bb
>> x = sym('x');         The symbolic variable x is created but not displayed,
>>                       since a semicolon is typed at the end of the command.
```

The name of the symbolic object can be different than the name of the variable. For example:

```
>> g = sym('gamma')      The symbolic object is gamma, and
                         the name of the object is g.
g =
gamma
```

As mentioned, symbolic objects can also be numbers. The numbers don't have to be typed as strings. For example, the `sym` command is used next to create symbolic objects from the numbers 5 and 7 and assign them to the variables *c* and *d*, respectively.

```
>> c = sym(5)    Create a symbolic object from the number 5 and assign it to c.

c =

5                        The display of a symbolic
>> d = sym(7)            object is not indented.

d =

7
```

As shown, when a symbolic object is created and a semicolon is not typed at the end of the command, MATLAB displays the name of the object and the object itself in the next two lines. The display of symbolic objects starts at the beginning of the line and is not indented like the display of numerical variables. The difference is illustrated below where a numerical variable is created.

```
>> e = 13
```
| | 13 is assigned to e (numerical variable). |

```
e =
   13  ◄────────────────────────
```
| | The display of the value of a numerical variable is indented. |

Several symbolic variables can be created in one command by using the syms command which has the form:

> syms variable_name variable_name variable_name

The command creates symbolic objects that have the same name as the symbolic variables. For example, the variables *y*, *z*, and *d* can all be created as symbolic variables in one command by typing:

```
>> syms y z d
>> y
y =
y
```
| | The variables created by the syms command are not displayed automatically. Typing the name of the variable shows that the variable was created, |

When the syms command is executed, the variables it creates are not displayed automatically – even if a semicolon is not typed at the end of the command.

11.1.2 Creating Symbolic Expressions

Symbolic expressions are mathematical expressions written in terms of symbolic variables. Once symbolic variables are created, they can be used for creating symbolic expressions. The symbolic expression is a symbolic object (the display is not indented). The form for creating a symbolic expression is:

> Expression_name = Mathematical expression

A few examples are:

```
>> syms a b c x y
```
| | Define a, b, c, x, and y as symbolic variables. |

```
>> f = a*x^2 + b*x + c
f =
a*x^2+b*x+c
      ◄──────────
```
| | Create the symbolic expression $ax^2 + bx + c$ and assign it to f. |
| | The display of the symbolic expression is not indented. |

When a symbolic expression, which includes mathematical operations that can be executed (addition, subtraction, multiplication, and division), is entered, MATLAB executes the operations as the expression is created. For example:

```
>> g = 2*a/3 + 4*a/7 - 6.5*x + x/3 + 4*5/3 - 1.5
```
| | $\dfrac{2a}{3} + \dfrac{4a}{7} - 6.5x + \dfrac{x}{3} + 4\dfrac{5}{3} - 1.5$ is entered. |

g =
26/21*a-37/6*x+31/6

$\dfrac{26}{21}a - \dfrac{37}{6}x + \dfrac{31}{6}$ is displayed.

Notice that all the calculations are carried out exactly with no numerical approximation. In the last example, $\dfrac{2}{3}a$ and $\dfrac{4}{7}a$ were added by MATLAB to give $\dfrac{26}{21}a$, and $-6.5x + \dfrac{x}{3}$ was added to $\dfrac{37}{6}x$. The operations with the terms that contain only numbers in the symbolic expression are carried out exactly. In the last example, $4 \cdot \dfrac{5}{3} + 1.5$ is replaced by $\dfrac{31}{6}$.

The difference between exact and approximate calculations is demonstrated in the following example, where the same mathematical operations are carried out once with symbolic variables and once with numerical variables.

```
>> a = sym(3); b = sym(5);
```
Define a and b as symbolic 3 and 5, respectively.

```
>> e = b/a + sqrt(2)
```
Create an expression that includes a and b.

```
e =
5/3+2^(1/2)
```
An exact value of e is displayed as a symbolic object (The display is not indented).

```
>> c = 3; d = 5;
```
Define c and d as numerical 3 and 5, respectively.

```
>> f = d/c + sqrt(2)
```
Create an expression that includes c and d.

```
f =
   3.0809
```
An approximated value of f is displayed as a number (The display is indented).

An expression that is created can include both symbolic objects and numerical variables. However, if an expression includes a symbolic object (one or several) all the mathematical operations will be carried out exactly. For example, if c is replaced by a in the last expression, the result is exact as it was in the first example.

```
>> g = d/a + sqrt(2)

g =
5/3+2^(1/2)
```

Additional facts about symbolic expressions and symbolic objects:

* Symbolic expressions can include numerical variables which have been obtained from the execution of numerical expressions. When these variables are inserted in symbolic expressions their exact value is used, even if the variable was displayed before with an approximated value. For example:

```
>> h = 10/3
```
h is defined to be 10/3 (a numerical variable).

h =
   3.3333   An approximated value of h (numerical variable) is displayed.

>> k = sym(5); m = sym(7);   Define k and m as symbolic 5 and 7, respectively.

>> p = k/m + h   h, k, and m are used in an expression.

p =
85/21   The exact value of h is used in the determination of p.
                An exact value of p (symbolic object) is displayed.

- The `double(S)` command can be used to convert a symbolic expression (object) S which is written in an exact form to numerical form. (The name double comes from the fact that the command returns a double-precision floating-point number representing the value of S). Two examples are shown. In the first, the p from the last example is converted into a numerical form. In the second, a symbolic object is created, and then converted into numerical form.

>> pN=double(p)   p is converted to numerical form (assigned to pN).

pN =
   4.0476

>> y = sym(10)*cos(5*pi/6)   Create a symbolic expression y.

y =
-5*3^(1/2)   Exact value of y is displayed.

>> yN=double(y)   y is converted to numerical form (assigned to yN).

yN =
   -8.6603

- A symbolic object that is created can also be a symbolic expression written in terms of variables that have not been first created as symbolic objects. For example, the quadratic expression $ax^2 + bx + c$ can be created as a symbolic object named f by using the `sym` command:

>> f = sym('a*x^2 + b*x +c')

f =
a*x^2 + b*x +c

It is important to understand that in this case, the variables a, b, c, and x included in the object do not exist individually as independent symbolic objects (the whole expression is one object). This means that it is impossible to perform symbolic math operations associated with the individual variables in the object. For example, it will not be possible to differentiate f with respect to x. This is different than the way in which the quadratic expression was created in the first example in Section *11.1.2*, where the individual variables are first created as symbolic objects and then used in the quadratic expression.

- Existing symbolic expressions can be used to create new symbolic expressions. This is done by simply using the name of the existing expression in the new expression. For example:

>> syms x y	Define x, and y as symbolic variables.
>> SA = x + y, SB = x - y	Create two symbolic expressions SA and SB.
SA =	
x+y	
SB =	$SA = x + y$
x-y	$SB = x - y$
>> F = SA^2/SB^3 + x^2	Create a new symbolic expression F using SA and SB.
F =	
(x+y)^2/(x-y)^3+x^2	$F = (SA^2)/(SB^3) + x^2 = \dfrac{(x+y)^2}{(x-y)^3} + x^2$

11.1.3 The `findsym` Command and the Default Symbolic Variable

The `findsym` command can be used to find which symbolic variables are present in an existing symbolic expression. The format of the command is:

$$\boxed{\texttt{findsym(S)}} \quad \text{or} \quad \boxed{\texttt{findsym(S,n)}}$$

The `findsym(S)` command displays the names of all the symbolic variables (separated by commas) that are in the expression S in alphabetical order. The `findsym(S,n)` command displays n symbolic variables which are in expression S in the default order. For one-letter symbolic variables, the default order starts with x and is followed by letters according to their closeness to x. If there are two letters that are equally close to x, the letter that is after x in the alphabetical order is first (y before w, and z before v). The default symbolic variable in a symbolic expression is the first variable in the default order. The default symbolic variable in an expression S can be identified by typing `findsym(S,1)`. Examples:

>> syms x h w y d t	Define x, h, w, y, d, and t as symbolic variables.
>> S = h*x^2 + d*y^2 + t*w^2	Create a symbolic expression S.
S =	
h*x^2+d*y^2+t*w^2	
>> findsym(S)	Use the `findsym(S)` command.
ans =	
d, h, t, w, x, y	The symbolic variables are displayed in alphabetical order.
>> findsym(S,5)	Use the `findsym(S,n)` command (n = 5).

ans =

x,y,w,t,h

> 5 symbolic variables are displayed in the default order.

>> findsym(S,1)

> Use the `findsym(S,n)` command with n = 1.

ans =

x

> The default symbolic variable is displayed.

11.2 CHANGING THE FORM OF AN EXISTING SYMBOLIC EXPRESSION

Symbolic expressions are either created by the user or by MATLAB as the result of symbolic operations. The expressions created by MATLAB might not be in the simplest form or in a form that the user prefers. The form of an existing symbolic expression can be changed by collecting terms with the same power, by expanding products, by factoring out common multipliers, by using mathematical and trigonometric identities, and by many other operations. The following sub-sections describe several of the commands that can be used to change the form of an existing symbolic expression.

11.2.1 The `collect`, `expand`, and `factor` Commands

The `collect`, `expand`, and `factor` commands can be used to perform the mathematical operations that are implied by their name.

The `collect` command:

The `collect` command collects the terms in the expression that have the variable with the same power. In the new expression, the terms will be ordered in decreasing order of power. The command has the forms:

collect(S)   collect(S, variable_name)

where S is the expression. The `collect(S)` form works best when an expression has only one symbolic variable. If an expression has more than one variable MATLAB will collect the terms of one variable first, then of a second variable, and so on. The order of the variables is determined by MATLAB. The user can specify the first variable by using the `collect(S, variable_name)` form of the command. Examples:

>> syms x y

> Define x and y as symbolic variables.

>> S = (x^2 + x - exp(x))*(x + 3)

S =

(x^2 + x-exp(x))*(x + 3)

> Create the symbolic expression S:
> $(x^2 + x - e^x)(x + 3)$.

>> F = collect(S)

> Use the `collect` command.

F =

x^3 + 4*x^2 + (-exp(x) + 3)*x - 3*exp(x)

> MATLAB returns the expression:
> $x^3 + 4x^2 + (-e^x + 3)x - 3e^x$.

>> T = (2*x^2 + y^2)*(x + y^2 + 3)
T =
(2*x^2+y^2)*(x+y^2+3)

Create the symbolic expression T:
$(2x^2 + y^2)(x + y^2 + 3)$.

>> G = collect(T)

Use the `collect(T)` command.

G =
2*x^3+(2*y^2+6)*x^2+x*y^2+y^2*(y^2+3)

MATLAB returns the expression:
$2x^3 + (2y^2 + 6)x^2 + xy^2 + y^2(y^2 + 3)$

>> H = collect(T,y)

Use the `collect(T,y)` command.

H =
y^4+(2*x^2+x+3)*y^2+2*x^2*(x+3)

MATLAB returns the expression:
$y^4 + (2x^2 + x + 3)y^2 + 2x^2(x + 3)$.

Note above that when `collect(T)` is used, the reformatted expression is written in terms of reduced powers of x, but when the `collect(T,y)` is used the reformatted expression is written in terms of reduced powers of y.

The `expand` command:

The `expand` command expands expressions in two ways. It carries out products of terms that include summation (at least one of the terms), and it uses trigonometric identities and exponential and logarithmic laws to expand corresponding terms that include summation. The form of the command is:

expand(S)

where S is the symbolic expression. Two examples are:

>> syms a x y

Define a, x, and y as symbolic variables.

>> S=(x + 5)*(x - a)*(x + 4)
S =
(x+5)*(x-a)*(x+4)

Create the symbolic expression S:
$(x + 5)(x - a)(x + 4)$.

>> T=expand(S)

Use the expand command.

T =
x^3+9*x^2-x^2*a-9*x*a+20*x-20*a

MATLAB returns the expression:
$x^3 + 9x^2 - ax^2 - 9ax + 20x - 20a$.

>> expand(sin(x-y))

Use the expand command to expand $\sin(x - y)$.

ans =
sin(x)*cos(y)-cos(x)*sin(y)

MATLAB uses trig identity for the expansion.

The `factor` command:

The `factor` command changes an expression that is a polynomial to be a product of polynomials of a lower degree. The form of the command is:

factor(S)

where S is the symbolic expression. An example is:

`>> syms x`	Define x as a symbolic variable.

`>> S = x^3 + 4*x^2 - 11*x - 30`
`S =`
`x^3+4*x^2-11*x-30`
	Create the symbolic expression S:
	$x^3 + 4x^2 - 11x - 30$.

`>> factor(S)`	Use the `factor` command.
`ans =` `(x+2)*(x-3)*(x+5)`	MATLAB returns the expression: $(x+2)(x-3)(x+5)$.

11.2.2 *The* `simplify` *and* `simple` *Commands*

The `simplify` and `simple` commands are both general tools for simplifying the form of an expression. The `simplify` command uses built-in simplification rules to generate a form of the expression with a simpler form than the original expression. The `simple` command is programmed to generate a form of the expression with the least number of characters. Although there is no guarantee that the form with the least number of characters is the simplest, in actuality this is often the case.

The `simplify` **command:**

The `simplify` command uses mathematical operations (addition, multiplication, rules of fractions, powers, logarithms, etc.) and functional and trigonometric identities to generate a simpler form of the expression. The format of the `simplify` command is:

$$\boxed{\texttt{simplify(S)}}$$

S is the name of the existing   or   An expression to be simplified
expression to be simplified.   can be typed in for S.

Two examples are:

`>> syms x y`	Define x as a symbolic variable.

`>> S = x*(x*(x - 8) + 10) - 5`
`S =`
`x*(x*(x - 8) + 10) - 5`
	Create the symbolic expression:
	$x(x(x-8)+10)-5$), and assign it to S.

`>> SA = simplify(S)`	Use the `simplify` command to Simplify S.
`SA =` `x^3-8*x^2+10*x-5`	MATLAB simplifies the expression to: $x^3 - 8x^2 + 10x - 5$.
`>> simplify((x + y)/(1/x + 1/y))`	Simplify $(x+y)/\left(\dfrac{1}{x}+\dfrac{1}{y}\right)$.
`ans =` `x*y`	MATLAB simplifies the expression to: xy).

The `simple` command:

The `simple` command finds a form of the expression with the fewest number of characters. In many cases this form is also the simplest. When the command is executed, MATLAB creates several forms of the expression by applying the `collect`, `expand`, `factor`, `simplify` commands, and other simplification functions that are not covered here. Then MATLAB returns the expression with the shortest form. The simple command has the following three forms:

F = simple(S)	simple(S)	[F how] = simple(S)
The shortest form of S is assigned to F.	All the simplification trails are displayed. The shortest is assigned to `ans`.	The shortest form of S is assigned to F. The name (string) of the simplification method is assigned to `how`.

The difference between the forms is in the output. The use of two of the forms is shown next.

```
>> syms x                                    Define x as a symbolic variable.

>> S = (x^3 - 4*x^2 + 16*x)/(x^3 + 64)       Create the symbolic expression:
S =
(x^3-4*x^2+16*x)/(x^3+64)                     x³ – 4x² + 16x
                                             ─────────────, and assign it to S.
                                                x³ + 64

>> F = simple(S)                             Use the F = simple(S) command to Simplify S.
F =
x/(x+4)                                       The simplest form of S, x/(x+4) is assigned to F.

>> [G how] = simple(S)                        Use the [G how] = simple(S) command.
G =
x/(x+4)                                       The simplest form of S, x/(x+4) is assigned to G.

how =                                         The word factor is assigned to G, which means that the
factor                                        shortest form was obtained using the factor command.
```

The use of the `simple(S)` form of the command is not demonstrated because the display of the output is lengthy. MATLAB displays ten different tries and assigns the shortest form to `ans`. The reader should try to execute the command and examine the output display.

11.2.3 The `pretty` Command

The `pretty` command displays a symbolic expression in a format resembling the mathematical format in which expressions are generally typed. The command has the form:

$$\boxed{\texttt{pretty(S)}}$$

For example:

>> syms a b c x | Define a, b, c, and x as symbolic variables.

>> S = sqrt(a*x^2 + b*x + c) | Create the symbolic expression:

S = | $\sqrt{ax^2 + bx + c}$, and assign it to S.
(a*x^2+b*x+c)^(1/2)

>> pretty(S) | The pretty command displays
                                  | the expression in a math format.

$$(a x^2 + b x + c)^{1/2}$$

11.3 SOLVING ALGEBRAIC EQUATIONS

A single algebraic equation can be solved for one variable, and a system of equations can be solved for several variables with the solve function.

Solving a single equation:

An algebraic equation can have one or several symbolic variables. If the equation has one variable, the solution is numerical. If the equation has several symbolic variables, a solution can be obtained for any of the variables in terms of the others. The solution is obtained by using the solve command which has the forms:

$$\boxed{\texttt{h = solve(eq)}} \quad \text{or} \quad \boxed{\texttt{h = solve(eq,var)}}$$

- The argument eq can be the name of a previously created symbolic expression, or an expression that is typed in. When a previously created symbolic expression S is entered for eq, or when an expression that does not contain the = sign is typed in for eq, MATLAB solves the equation eq = 0.

- An equation of the form $f(x) = g(x)$ can be solved by typing the equation (including the = sign) as a string for eq.

- If the equation to be solved has more than one variable, the solve(eq) command solves for the default symbolic variable (Section *11.1.3*). A solution for any of the variables can be obtained with the solve(eq,var) command by typing the variable name for var.

- If the user types: solve(eq), the solution is assigned to the variable ans.

- If the equation has more than one solution, the output h is a symbolic column vector with a solution at each element. The elements of the vector are symbolic objects. When an array of symbolic objects is displayed, each row is enclosed with square brackets (see the examples on the following page).

The following examples illustrate the use of the `solve` command.

>> syms a b x y z | Define a, b, x, y and z as symbolic variables.

>> h = solve(exp(2*z) - 5) | Use the `solve` command to solve $e^{2z} - 5 = 0$.

h =

1/2*log(5) | The solution is assigned to h.

>> S = x^2 - x - 6 | Create the symbolic expression:
S = | $x^2 - x - 6$, and assign it to S.
x^2-x-6

>> k = solve(S) | Use the `solve(S)` command to solve $x^2 - x - 6 = 0$.

k = | The equation has two solutions. They are assigned to
[-2] | k, which is a column vector with symbolic objects.
[3]

>> solve('cos(2*y) + 3*sin(y) = 2') | Use the `solve` command to solve:
                                       | $\cos(2y) + 3\sin(y) = 2$. (The equation is
ans = | typed as a string in the command.)
[1/2*pi]
[1/6*pi]
[5/6*pi] | The solution is assigned to ans.

>> T = a*x^2 + 5*b*x + 20 | Create the symbolic expression:
T = | $ax^2 + 5bx + 20$, and assign it to T.
a*x^2+5*b*x+20

>> solve(T) | Use the `solve(S)` command to solve $T = 0$.

ans = | The equation $T = 0$ is solved for the
[1/2/a*(-5*b+(25*b^2-80*a)^(1/2))] | variable x, which is the default variable.
[1/2/a*(-5*b-(25*b^2-80*a)^(1/2))]

>> M = solve(T,a) | Use the `solve(eq,var)` command to solve $T = 0$.

M = | The equation $T = 0$ is solved for the variable a.
-5*(b*x+4)/x^2

- It is also possible to use the `solve` command by typing the equation to be solved as a string, without having the variables in the equation first created as symbolic objects. However, if the solution contains variables (when the equation has more than one variable), the variables do not exist as independent symbolic objects. For example:

>> ts = solve('4*t*h^2 + 20*t - 5*g') | The expression $4th^2 + 20t - 5g$ is
                                          | typed in the `solve` command.

                                          | The variables t, h, and g have not been created as symbolic vari-
ts = | ables before the expression was typed in the `solve` command.
5/4*g/(h^2+5) | MATLAB solves the equation $4th^2 + 20t - 5g = 0$ for t.

The equation can also be solved for a different variable. For example, a solution for g is obtained by:

```
>> gs = solve('4*t*h^2 + 20*t - 5*g','g')

gs =
4/5*t*h^2+4*t
```

Solving a system of equations:

The `solve` command can also be used for solving a system of equations. If the number of equations and variables is the same, the solution is numerical. If the number of variables is greater than the number of equations, the solution is symbolic for the desired variables in terms of the other variables. A system of equations (depending on the type of equations) can have one or several solutions. If the system has one solution, each of the variables for which the system is solved has one numerical value (or expression). If the system has more than one solution, each of the variables can have several values.

The format of the `solve` command for solving a system of n equations is:

$$\boxed{\texttt{output = solve(eq1,eq2,....,eqn)}}$$

or

$$\boxed{\texttt{output = solve(eq1,eq2,....,eqn,var1,var2,....varn)}}$$

- The arguments `eq1,eq2,....,eqn` are the equations to be solved. Each argument can be a name of a previously created symbolic expression, or an expression that is typed in as a string. When a previously created symbolic expression `S` is entered, the equation is `S = 0`. When a string that does not contain the `=` sign is typed in, the equation is `expression = 0`. An equation that contains the `=` sign must be typed as a string.

- In the first format, if the number of equations n is equal to the number of variables in the equations, MATLAB gives a numerical solution for all the variables. If the number of variables is greater than the number of equations n, MATLAB gives a solution for n variables in terms of the rest of the variables. The variables for which solutions are obtained are chosen by MATLAB according to the default order (Section *11.1.3*).

- When the number of variables is greater than the number of equations n, the user can select the variables for which the system is solved. This is done by using the second format of the `solve` command and entering the names of the variables `var1,var2,....varn`.

The `output` from the `solve` command, which is the solution of the system, can have two different forms. One is a cell array and the other is a structure. A cell array is an array in which each of the elements can be an array. A structure is an array in which the elements (called fields) are addressed by textual field designators. The fields of a structure can be arrays of different sizes and types. Cell arrays and structures are not presented in detail in this book, but a short explanation is given below, so the reader will be able to use them with the `solve` command.

When a cell array is used in the output of the `solve` command, the command has the following form (in the case of a system of three equations):

$$[varA, varB, varC] = solve(eq1, eq2, eq3)$$

- Once the command is executed, the solution is assigned to the variables `varA`, `vary`, `varC`, and the variables are displayed with their assigned solution. Each of the variables will have one or several values (in a column vector) depending whether the system of equations has one or several solutions.

- The user can select any name for `varA`, `varB`, `varC`. MATLAB assigns the solution for the variables in the equations in alphabetical order. For example, if the variables for which the equations are solved are x, u, and t, the solution for t is assigned to `varA`, the solution for u is assigned to `varB`, and the solution for x is assigned to `varC`.

The following examples show how the `solve` command is used for the case when cell array is used in the output:

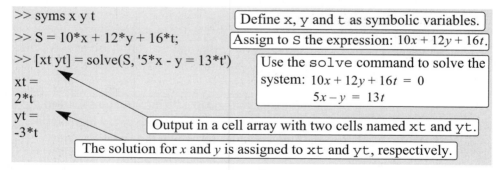

```
>> syms x y t                          Define x, y and t as symbolic variables.
>> S = 10*x + 12*y + 16*t;             Assign to S the expression: 10x + 12y + 16t.
>> [xt yt] = solve(S, '5*x - y = 13*t')   Use the solve command to solve the
xt =                                      system:  10x + 12y + 16t = 0
2*t                                                5x - y = 13t
yt =
-3*t                   Output in a cell array with two cells named xt and yt.
          The solution for x and y is assigned to xt and yt, respectively.
```

In the example above, notice that the system of two equations is solved by MATLAB for x and y in terms of t, since x and y are the first two variables in the default order. The system, however, can be solved for different variables. As an example, the system is solved next for y and t in terms of x (using the second form of the `solve` command:

```
>> [tx yx] = solve(S, '5*x - y = 13*t', y, t)
                                    The variables for which the system
                                    is solved (y and t) are entered.
```

tx =
1/2*x
yx =
-3/2*x

> The solution for the variables for which the system is solved is assigned in alphabetical order. The first cell has the solution for *t*, and the second cell has the solution for *y*.

When a structure is used in the output of the `solve` command, the command has the form (in the case of a system of three equations):

$$AN = \texttt{solve(eq1,eq2,eq3)}$$

- `AN` is the name of the structure.

- Once the command is executed the solution is assigned to `AN`. MATLAB displays the name of the structure and the names of the fields of the structure which are the names of the variables for which the equations are solved. The size and the type of each field is displayed next to the field name. The content of each field, which is the solution for the variable, is not displayed.

- To display the content of a field (the solution for the variable), the user has to type the address of the field. The form for typing the address is: `structure_name.field_name` (see example below).

As an example, the system of equations solved in the last example is solved again using a structure for the output.

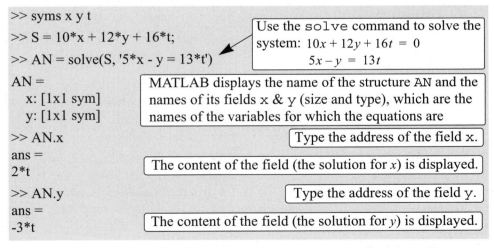

>> syms x y t
>> S = 10*x + 12*y + 16*t;
>> AN = solve(S, '5*x - y = 13*t')

> Use the `solve` command to solve the system: $10x + 12y + 16t = 0$
> $5x - y = 13t$

AN =
x: [1x1 sym]
y: [1x1 sym]

> MATLAB displays the name of the structure AN and the names of its fields x & y (size and type), which are the names of the variables for which the equations are

>> AN.x

> Type the address of the field x.

ans =
2*t

> The content of the field (the solution for *x*) is displayed.

>> AN.y

> Type the address of the field y.

ans =
-3*t

> The content of the field (the solution for *y*) is displayed.

Sample problem 11-1 shows a solution of a system of equations that have two solutions.

Sample Problem 11-1: Intersection between a circle and a line

The equation of a circle in the x-y plane with a radius R and its center at point (2, 4) is given by: $(x-2)^2 + (y-4)^2 = R^2$. The equation of a line in the plane is given by: $y = \frac{x}{2} + 1$. Determine the coordinates of the points (as a function of R) where the line intersects the circle.

Solution

The solution is obtained by solving the system of the two equations for x and y in terms of R. To show the difference in the output between using cell array and structure output forms of the `solve` command, the system is solved twice. The first solution is with the output in a cell array:

```
>> syms x y R
```
The two equations are typed in the `solve` command.

```
>> [xc, yc] = solve('(x - 2)^2 + (y - 4)^2 = R^2', 'y = x/2 + 1')
```
Output in a cell array.

```
xc =
[ 14/5+2/5*(-16+5*R^2)^(1/2)]
[ 14/5-2/5*(-16+5*R^2)^(1/2)]
yc =
[ 12/5+1/5*(-16+5*R^2)^(1/2)]
[ 12/5-1/5*(-16+5*R^2)^(1/2)]
```
Output in a cell array with two cells named `xc` and `yc`. Each cell contains two solutions in a symbolic column vector.

The second solution is with the output in a structure:

```
>> COORD= solve('(x - 2)^2 + (y - 4)^2 = R^2', 'y = x/2 + 1')
```
Output in a structure.

```
COORD =
  x: [2x1 sym]
  y: [2x1 sym]
```
Output in a structure named `COORD` that has two fields `x` and `y`. Each field is a 2 by 1 symbolic vector.

```
>> COORD.x
```
Type the address of the field `x`.

```
ans =
[ 14/5+2/5*(-16+5*R^2)^(1/2)]
[ 14/5-2/5*(-16+5*R^2)^(1/2)]
```
The content of the field (the solution for x) is displayed.

```
>> COORD.y
```
Type the address of the field `y`.

```
ans =
[ 12/5+1/5*(-16+5*R^2)^(1/2)]
[ 12/5-1/5*(-16+5*R^2)^(1/2)]
```
The content of the field (the solution for y) is displayed.

11.4 DIFFERENTIATION

Symbolic differentiation can be carried out by using the `diff` command. The form of the command is:

$$\boxed{\texttt{diff(S)}} \quad \text{or} \quad \boxed{\texttt{diff(S,var)}}$$

- `S` can either be a name of a previously created symbolic expression, or an expression can be typed in for `S`.

- In the `diff(S)` command, if the expression contains one symbolic variable, the differentiation is carried out with respect to that variable. If the expression contains more than one variable, the differentiation is carried out with respect to the default symbolic variable (Section *11.1.3*).

- In the `diff(S,var)` command (which is used for differentiation of expressions with several symbolic variables) the differentiation is carried out with respect to the variable `var`.

- Second or higher (*n*-th) derivative can determined with the `diff(S,n)` or `diff(S,var,n)` command, where n is a positive number. $n = 2$ for the second derivative, 3 for the third, and so on.

Some examples are:

```
>> syms x y t
```
Define x, y and t as symbolic variables.

```
>> S = exp(x^4);
```
Assign to S the expression: e^{x^4}.

```
>> diff(S)
```
Use the `diff(S)` command to differentiate S.

```
ans =
4*x^3*exp(x^4)
```
The answer $4x^3e^{x^4}$ is displayed.

```
>> diff((1 - 4*x)^3)
```
Use the `diff(S)` command to differentiate $(1 - 4x)^3$.

```
ans =
-12*(1-4*x)^2
```
The answer $-12(1-4x)^2$ is displayed.

```
>> R = 5*y^2*cos(3*t);
```
Assign to R the expression: $5y^2\cos(3t)$.

```
>> diff(R)
```
Use the `diff(R)` command to differentiate R.

```
ans =
10*y*cos(3*t)
```
MATLAB differentiates R with respect to y (default symbolic variable), the answer $10y\cos(3t)$ is displayed.

```
>> diff(R,t)
```
Use the `diff(R,t)` command to differentiate R w.r.t. *t*.

```
ans =
-15*y^2*sin(3*t)
```
The answer $-15y^2\sin(3t)$ is displayed.

```
>> diff(S,2)
```
Use `diff(S,2)` command to obtain the 2nd derivative of S.

```
ans =
12*x^2*exp(x^4)+16*x^6*exp(x^4)
```
The answer $12x^2e^{x^4} + 16x^6e^{x^4}$ is displayed.

- It is possible also to use the `diff` command by typing the expression to be differentiated as a string directly in the command without having the variables in the expression first created as symbolic objects. However, the variables in the differentiated expression do not exist as independent symbolic objects.

11.5 INTEGRATION

Symbolic integration can be carried out by using the `int` command. The command can be used for determining indefinite integrals (antiderivatives) and definite integrals. For indefinite integration the form of the command is:

$$\boxed{\texttt{int(S)}} \quad \text{or} \quad \boxed{\texttt{int(S,var)}}$$

- `S` can either be a name of a previously created symbolic expression, or an expression can be typed in for `S`.

- In the `int(S)` command, if the expression contains one symbolic variable, the integration is carried out with respect to that variable. If the expression contains more than one variable, the integration is carried out with respect to the default symbolic variable (Section *11.1.3*).

- In the `int(S,var)` command, which is used for integration of expressions with several symbolic variables, the integration is carried out with respect to the variable `var`.

Some examples are:

`>> syms x y t`	Define `x`, `y` and `t` as symbolic variables.
`>> S = 2*cos(x) - 6*x;`	Assign to `S` the expression: $2\cos(x) - 6x$.
`>> int(S)`	Use the `int(S)` command to integrate `S`.
`ans =` `2*sin(x)-3*x^2`	The answer $2\sin(x) - 3x^2$ is displayed.
`>> int(x*sin(x))`	Use the `int(S)` command to integrate $x\sin(x)$.
`ans =` `sin(x)-x*cos(x)`	The answer $\sin(x) - x\cos(x)$ is displayed.
`>>R = 5*y^2*cos(4*t);`	Assign to `R` the expression: $5y^2\cos(4t)$.
`>> int(R)`	Use the `int(R)` command to integrate `R`.
`ans =` `5/3*y^3*cos(4*t)`	MATLAB integrate `R` with respect to y (default symbolic variable), the answer $5y^3\cos(4t)/3$ is displayed.
`>> int(R,t)`	Use the `int(R,t)` command to integrate `R` w.r.t. t.
`ans =` `5/4*y^2*sin(4*t)`	The answer $5y^2\sin(4t)/4$ is displayed.

For definite integration the form of the command is:

$$\boxed{\texttt{int(S,a,b)}} \quad \text{or} \quad \boxed{\texttt{int(S,var,a,b)}}$$

- `a` and `b` are the limits of integration. The limits can be numbers or symbolic variables.

For example, determination of the definite integral $\int_0^\pi (\sin y + 5y^2)dy$ with MAT-

LAB is:

```
>> syms y
>> int(sin(y) - 5*y^2, 0, pi)

ans =
2-5/3*pi^3
```

- It is possible also to use the `int` command by typing the expression to be integrated as a string without having the variables in the expression first created as symbolic objects. However, the variables in the integrated expression do not exist as independent symbolic objects.

- Integration can sometimes be a difficult task. A closed form answer may not exist, or if exists, MATLAB might not be able to find it. When that happens MATLAB returns `int(S)` and a message `Explicit integral could not be found`.

11.6 SOLVING AN ORDINARY DIFFERENTIAL EQUATION

An ordinary differential equation (ODE) can be solved symbolically with the `dsolve` command. The command can be used to solve a single equation or a system of equations. Only single equations are addressed here. Chapter 10 discusses using MATLAB to solve first order ODE's numerically. The reader's familiarity with the subject of differential equations is assumed. The purpose of this section is to show how to use MATLAB for solving such equations.

A first order ODE is an equation that contains the derivative of the dependent variable. If t is the independent variable and y is the dependent variable, the equation can be written in the form:

$$\frac{dy}{dt} = f(t, y)$$

A second order ODE contains the second derivative of the dependent variable (it can also contain the first derivative). Its general form is:

$$\frac{d^2y}{dt^2} = f\left(t, y, \frac{dy}{dt}\right)$$

A solution is a function $y = f(t)$ that satisfies the equation. The solution can be general or particular. A general solution contains constants. In a particular solution the constants are determined to have specific numerical values such that the solution satisfies specific initial or boundary conditions.

The command `dsolve` can be used for obtaining a general solution, or, when the initial or boundary conditions are specified, for obtaining a particular solution.

General solution:

For obtaining a general solution, the `dsolve` command has the form:

$$\boxed{\texttt{dsolve('eq')}} \qquad \text{or} \qquad \boxed{\texttt{dsolve('eq','var')}}$$

- `eq` is the equation to be solved. It has to be typed as a string (even if the variables are symbolic objects).

- The variables in the equation don't have to first be created as symbolic objects. (Then, in the solution the variables will not be symbolic objects.)

- Any letter (lower or upper case), except D can be used for the dependent variable.

- In the `dsolve('eq')` command the independent variable is assumed (default) by MATLAB to be `t`.

- In the `dsolve('eq','var')` command the user defines the independent variable by typing it for `var` (as a string).

- When typing the equation the letter D denotes differentiation. If y is the dependent variable and t is the independent variable, Dy stands for $\frac{dy}{dt}$. For example, the equation $\frac{dy}{dt} + 3y = 100$ is typed in as: `'Dy + 3*y = 100'`.

- A second derivative is typed by D2, third derivative by D3, and so on. For example, the equation $\frac{d^2y}{dt^2} + 3\frac{dy}{dt} + 5y = \sin(t)$ is typed in as: `'D2y + 3*Dy + 5*y = sin(t)'`.

- The variables in the ODE equation that is typed in the `dsolve` command do not have to be previously created symbolic variables.

- In the solution MATLAB uses C1, C2, C3, and so on, for the constants of integration.

For example, a general solution of the first order ODE: $\frac{dy}{dt} = 4t + 2y$ is obtained by:

```
>> dsolve('Dy = 4*t + 2*y')
```

ans =
-2*t-1+exp(2*t)*C1

> The answer $y = -2t - 1 + C_1 e^{2t}$ is displayed.

A general solution of the second order ODE: $\dfrac{d^2x}{dt^2} + 2\dfrac{dx}{dt} + x = 0$ is obtained by:

```
>> dsolve('D2x + 2*Dx + x = 0')
```

ans =
C1*exp(-t)+C2*exp(-t)*t

> The answer $x = C_1 e^{-t} + C_2 t e^{-t}$ is displayed.

The following examples illustrate the solution of differential equations that contain symbolic variables in addition to the independent and dependent variables.

```
>> dsolve('Ds = a*x^2')
```
> The independent variable is t (default).
> MATLAB solves the equation: $\dfrac{ds}{dt} = ax^2$.

ans =
a*x^2*t+C1

> The solution $s = ax^2 t + C_1$ is displayed.

```
>> dsolve('Ds = a*x^2','x')
```
> The independent variable is defined to be x.
> MATLAB solves the equation: $\dfrac{ds}{dx} = ax^2$.

ans =
1/3*a*x^3+C1

> The solution $s = \dfrac{1}{3}ax^3 + C_1$ is displayed.

```
>> dsolve('Ds = a*x^2','a')
```
> The independent variable is defined to be a.
> MATLAB solves the equation: $\dfrac{ds}{da} = ax^2$.

ans =
1/2*a^2*x^2+C1

> The solution $s = \dfrac{1}{2}a^2 x^2 + C_1$ is displayed.

Particular solution:

A particular solution of an ODE can be obtained if boundary (or initial) conditions are specified. A first order equation requires one condition, a second order equation requires two conditions, and so on. For obtaining a particular solution, the dsolve command has the form:

First order ODE:
```
dsolve('eq','cond1','var')
```

Higher order ODE:
```
dsolve('eq','cond1','cond2',....,'var')
```

- For solving equations of higher order, additional boundary conditions have to be entered in the command. If the number of conditions is less than the order of the equation, MATLAB returns a solution that includes constants of integration (C1, C2, C3, and so on).

- The boundary conditions are typed in as strings in the following way:

Math form	MATLAB form
$y(a) = A$	`'y(a)=A'`
$y'(a) = A$	`'Dy(a)=A'`
$y''(a) = A$	`'D2y(a)=A'`

- The argument `'var'` is optional, and is used to define the independent variable in the equation. If none is entered, the default is t.

For example, the first order ODE: $\dfrac{dy}{dt} + 4y = 60$, with the initial condition: $y(0) = 5$ is solved with MATLAB by:

```
>> dsolve('Dy + 4*y = 60','y(0) = 5')

ans =
15-10*exp(-4*t)
```
The answer $y = 15 - 10e^{-4t}$ is displayed.

The second order ODE: $\dfrac{d^2y}{dt^2} - 2\dfrac{dy}{dt} + 2y = 0$, $y(0) = 1$, $\left.\dfrac{dy}{dt}\right|_{t=0} = 0$ can be solved with MATLAB by:

```
>> dsolve('D2y - 2*Dy + 2*y = 0','y(0) = 1','Dy(0) = 0')

ans =
-exp(t)*sin(t)+exp(t)*cos(t)
```
The answer $y = -e^t \sin(t) + e^t \cos(t)$ is displayed.

```
>> factor(ans)
```
The answer can be simplified with the `factor` command.

```
ans =
-exp(t)*(sin(t)-cos(t))
```
The simplified answer $y = -e^t(\sin(t) - \cos(t))$ is displayed.

Additional examples of solving differential equations are shown in Sample Problem 11-5.

If MATLAB cannot find a solution, it returns an empty symbolic object, and a message: `Warning: explicit solution could not be found`.

11.7 PLOTTING SYMBOLIC EXPRESSIONS

In many cases, there is a need to plot a symbolic expression. This can easily be done with the `ezplot` command. For a symbolic expression S that contains one variable `var`, MATLAB considers the expression to be a function $S(var)$, and the command creates a plot of $S(var)$ versus var. For a symbolic expression that contains two symbolic variables `var1` and `var2`, MATLAB considers the expression to be a function in the form $S(var1, var2) = 0$, and the command creates a plot of one variable versus the other.

To plot a symbolic expression S that contains one or two variables, the `ezplot` command is:

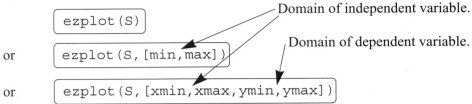

- `S` is the symbolic expression to be plotted. It can be the name of a previously created symbolic expression, or an expression can be typed in for `S`.

- It is also possible to type the expression to be plotted as a string without having the variables in the expression first created as symbolic objects.

- If `S` has one symbolic variable, a plot of $S(var)$ versus (var) is created, with the values of *var* (the independent variable) on the abscissa (horizontal axis), and the values of the $S(var)$ on the ordinate (vertical axis).

- If the symbolic expression `S` has two symbolic variables, `var1` and `var2`, the expression is assumed to be a function with the form $S(var1,var2) = 0$. MAT-LAB creates a plot of one variable versus the other variable. The variable that is first in alphabetic order is taken to be the independent variable. For example, if the variables in *S* are *x* and *y*, then *x* is the independent variable and is plotted on the abscissa and *y* is the dependent variable plotted on the ordinate. If the variables in *S* are *u* and *v*, then *u* is the independent variable and *v* is the dependent variable.

- In the `ezplot(S)` command, if `S` has one variable ($S(var)$), the plot is over the domain: $-2\pi < var < 2\pi$ (default domain) and the range is selected by MATLAB. If `S` has two variables ($S(var1,var2)$), the plot is over $-2\pi < var1 < 2\pi$ and $-2\pi < var2 < 2\pi$.

- In the `ezplot(S,[min,max])` command the domain for the independent variable is defined by `min` and `max`: $min < var < max$ and the range is selected by MATLAB.

- In the `ezplot(S,[xmin,xmax,ymin,ymax]` command the domain for the independent variable is defined by `xmin` and `xmax`, and the domain of the dependent variable is defined by `ymin` and `ymax`.

The `ezplot` command can also be used to plot a function that is given in a parametric form. In this case two symbolic expressions, `S1` and `S2`, are involved, where each expression is written in terms of the same symbolic variable (independent parameter). For example, a plot of *y* vs. *x* where $x = x(t)$ and $y = y(t)$. In this case the form of the `ezplot` command is:

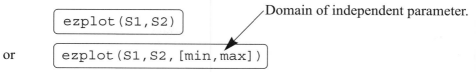

ezplot(S1,S2)

or ezplot(S1,S2,[min,max])

Domain of independent parameter.

- `S1` and `S2` are symbolic expressions containing the same single symbolic variable, which is the independent parameter. `S1` and `S2` can be the names of previously created symbolic expressions, or expressions can be typed in.

- The command creates a plot of $S2(var)$ versus $S1(var)$. The symbolic expression that is typed first in the command (`S1` in the definition above) is used for the horizontal axis and the expression that is typed second (`S2` in the definition above) is used for the vertical axis.

- In the `ezplot(S1,S2)` command the domain of the independent variable is: $0 < var < 2\pi$ (default domain).

- In the `ezplot(S1,S2,[min,max])` command the domain for the independent variable is defined by `min` and `max`: $min < var < max$.

Additional comments:

Once a plot is created, it can be formatted in the same way as plots created with the `plot` or `fplot` format. This can be done in two ways; by using commands, or by using the Plot Editor (see Section 5.4). When the plot is created, the expression that is plotted is displayed automatically at the top of the plot. MATLAB has additional easy plot functions for plotting two-dimensional polar plots, and for plotting three-dimensional plots. For more information, the reader is referred to the Help menu of the Symbolic Math Toolbox.

Several examples of using the `ezplot` command are shown in Table 11-1.

Table 11-1: Plots with the `ezplot` command

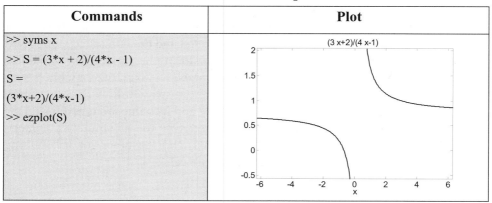

Table 11-1: Plots with the `ezplot` command (Continued)

Commands	Plot
>> syms x y >> S = 4*x^2 - 18*x + 4*y^2 + 12*y - 11 S = 4*x^2-18*x+4*y^2+12*y-11 >> ezplot(S)	
>> syms t >> x=cos(2*t) x = cos(2*t) >> y=sin(4*t) y = sin(4*t) >> ezplot(x,y)	

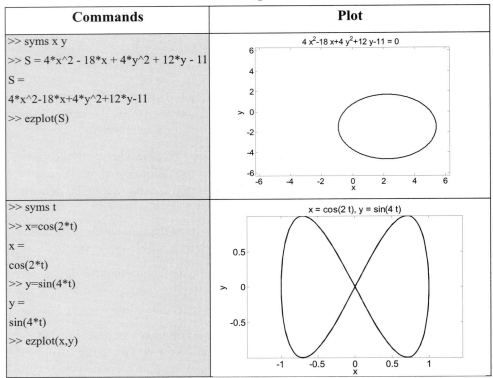

11.8 NUMERICAL CALCULATIONS WITH SYMBOLIC EXPRESSIONS

Once a symbolic expression is created by the user or by the output from any of MATLAB's symbolic operations, there might be a need to substitute numbers for the symbolic variables and calculate the numerical value of the expression. This can be done by using the `subs` command. The `subs` command has several forms and can be used in different ways. The following describes several forms that are easy to use and are suitable for most applications. In one form, the variable (or variables) for which a numerical value is substituted and the numerical value itself are typed inside the `subs` command. In another form, the variable (or variables) is assigned a numerical value in a separate command and then the variable is substituted in the expression.

The `subs` command in which the variable and its value are typed inside the command is shown first. Two cases are presented. One for substituting a numerical value (or values) for one symbolic variable, and the other for substituting numerical values for two or more symbolic variables.

Substituting a numerical value for one symbolic variable:

A numerical value (or values) can be substituted for one symbolic variable when a symbolic expression has one or more symbolic variables. In this case the `subs` command has the form:

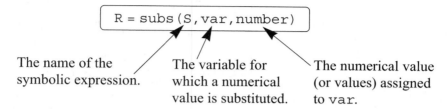

R = subs(S,var,number)

|The name of the symbolic expression.|The variable for which a numerical value is substituted.|The numerical value (or values) assigned to var.|

- `number` can be one number (scalar), or an array with many elements (a vector or a matrix).

- The value of S is calculated for each value of `number` and the result is assigned to R, which will have the same size as `number` (scalar, vector or matrix).

- If S has one variable, the output R is numerical. If S has several variables and a numerical value is substituted for only one of them, the output R is a symbolic expression.

An example with an expression that includes one symbolic variable is:

```
>> syms x
```
Define x, as a symbolic variable.
```
>> S = 0.8*x^3 + 4*exp(0.5*x)
S =
4/5*x^3+4*exp(1/2*x)
```
Assign to S the expression: $0.8x^3 + 4e^{(0.5x)}$.
```
>> SD = diff(S)
```
Use the `diff(S)` command to differentiate S.
```
SD =
12/5*x^2+2*exp(1/2*x)
```
The answer $12x^2/5 + 2e^{(0.5x)}$ is assigned to SD.
```
>> subs(SD, x, 2)
```
Use the `subs` command to substitute $x = 2$ in SD.
```
ans =
   15.0366
```
The value of SD is displayed.
```
>> SDU = subs(SD, x, [2:0.5:4])
```
Use the `subs` command to substitute $x = [2, 2.5, 3, 3.5, 4]$ (vector) in SD.
```
SDU =
   15.0366   21.9807   30.5634   40.9092   53.1781
```
The values of SD (assigned to SDU) for each value of x are displayed in a vector.

In the last example, notice that when the numerical value of the symbolic expression is calculated, the answer is numerical (the display is indented). An example of substituting numerical values for one symbolic variable in an expression that has several symbolic variables is:

>> syms a g t v   Define a, g, t and v as symbolic variables.

>> Y = v^2*exp(a*t)/g
                                   Create the symbolic expression:
Y = $v^2 e^{(at)}/g$ and assigned it to Y.
v^2*exp(a*t)/g

>> subs(Y, t, 2)   Use the subs command to substitute $t = 2$ in SD.

ans =   The answer $v^2 e^{(2a)}/g$ is displayed.
v^2*exp(2*a)/g

>> Yt = subs(Y, t, [2:4])   Use the subs command to substitute
                                   $t = [2, 3, 4]$ (vector) in Y.

Yt =
[v^2*exp(2*a)/g, v^2*exp(3*a)/g, v^2*exp(4*a)/g]

The answer is a vector with elements of symbolic expressions for each value of t.

Substituting a numerical value for two or more symbolic variables:

A numerical value (or values) can be substituted for two or more symbolic variables when a symbolic expression has several symbolic variables. In this case the subs command has the form (it is shown for two variables, but can be used in the same form for more):

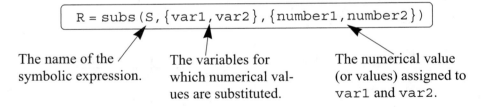

R = subs(S,{var1,var2},{number1,number2})

The name of the symbolic expression.   The variables for which numerical values are substituted.   The numerical value (or values) assigned to var1 and var2.

- The variables var1 and var2 are the variables in the expression S for which the numerical values are substituted. The variables are typed as a cell array (inside curly braces { }). Cell array is an array of cells where each cell can be an array of numbers or text.

- The numbers number1, number2 substituted for the variables are also typed as a cell array (inside curly braces { }). The numbers can be scalars, vectors, or matrices. The first cell in the numbers cell array (number1) is substituted for the variable that is in the first cell of the variable cell array (var1), and so on.

- If all the numbers that are substituted for variables are scalars, the outcome will be one number or one expression (if some of the variables are still symbolic).

- If, for at least one variable, the substituted numbers are an array, the mathematical operations are executed element-by-element and the outcome is an array of numbers or expressions. It should be emphasized that the calculations are performed element-by-element even though the expression S is not typed in the element-by-element notation. This also means that all the arrays substituted for different variables must be of the same size.

- It is possible to substitute arrays (of the same size) for some of the variables and scalars for other variables. In this case, in order to carry out element-by-element operations, MATLAB expands the scalars (array of ones times the scalar) to produce an array result.

The substitution of numerical values for two or more variables is demonstrated in the next examples.

```
>> syms a b c e x
```
Define a, b, c, e, and x as symbolic variables.

```
>> S= a*x^e + b*x + c
S =
a*x^e+b*x+c
```
Create the symbolic expression: $ax^e + bx + c$ and assigned it to S.

```
>> subs(S,{a,b,c,e,x},{5,4,-20,2,3})
```
Cell array. Cell array.

Substitute in S scalars for all the symbolic variables.

```
ans =
    37
```
The value of S is displayed.

```
>> T = subs(S,{a,b,c},{6,5,7})
T =
6*x^e+5*x+7
```
Substitute in S scalars for the symbolic variables a, b, and c.

The result is an expression with the variables x, and e.

```
>> R = subs(S,{b,c,e},{[2 4 6], 9, [1 3 5]})
```
Substitute in S a scalar for c, and vectors for b, and e.

```
R =
[  a*x+2*x+9, a*x^3+4*x+9, a*x^5+6*x+9]
```
The result is a vector of symbolic expressions.

```
>> W = subs(S,{a,b,c,e,x},{[4 2 0], [2 4 6], [2 2 2], [1 3 5], [3 2 1]})
```
Substitute in S vectors for all the variables.

```
W =
   20   26    8
```
The result is a vector of numerical values.

A second method to substitute numerical values for symbolic variables in a symbolic expression is first to assign numerical values to the variables, then to use the subs command. In this method, after the symbolic expression exists (at this point the variables in the expression are symbolic) the variables are assigned numerical values. Then, the subs command is used in the form:

R = subs (S)   The name of the
symbolic expression.

Once the symbolic variables are redefined as numerical variables they can no longer be used as symbolic. The method is demonstrated is the following examples.

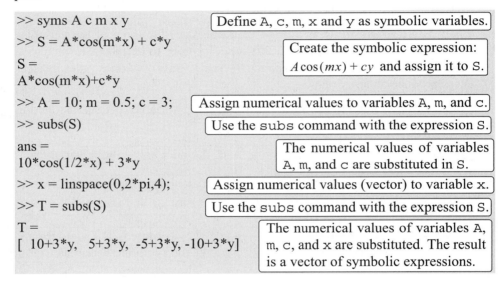

```
>> syms A c m x y
```
Define A, c, m, x and y as symbolic variables.
```
>> S = A*cos(m*x) + c*y
```
Create the symbolic expression:
$A\cos(mx) + cy$ and assign it to S.
```
S =
A*cos(m*x)+c*y
```
```
>> A = 10; m = 0.5; c = 3;
```
Assign numerical values to variables A, m, and c.
```
>> subs(S)
```
Use the subs command with the expression S.
```
ans =
10*cos(1/2*x) + 3*y
```
The numerical values of variables A, m, and c are substituted in S.
```
>> x = linspace(0,2*pi,4);
```
Assign numerical values (vector) to variable x.
```
>> T = subs(S)
```
Use the subs command with the expression S.
```
T =
[ 10+3*y,  5+3*y,  -5+3*y, -10+3*y]
```
The numerical values of variables A, m, c, and x are substituted. The result is a vector of symbolic expressions.

11.9 EXAMPLES OF MATLAB APPLICATIONS

Sample Problem 11-2: Firing angle of a projectile

A projectile is fired at a speed of 210 m/s and an angle θ. The projectile's intended target is 2600 m away and 210 m above the firing point.

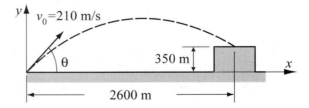

a) Derive the equation that has to be solved in order to determine the angle θ such that the projectile will hit the target.

b) Use MATLAB to solve the equation derived in part *a*.

c) For the angle determined in part *b*, use the ezplot command to make a plot of the projectile's trajectory.

Solution

a) The motion of the projectile can be analyzed by considering the horizontal and vertical components. The initial velocity v_0 can be resolved into horizontal and vertical components:

$$v_{0x} = v_0 \cos(\theta) \quad \text{and} \quad v_{0y} = v_0 \sin(\theta)$$

In the horizontal direction the velocity is constant, and the position of the projectile as a function of time is given by:

$$x = v_{0x}t$$

Substituting $x = 2600$ m for the horizontal distance that the projectile travels to the target, and $210\cos(\theta)$ for v_{0x}, and solving for t gives:

$$t = \frac{2600}{210\cos(\theta)}$$

In the vertical direction the position of the projectile is given by:

$$y = v_{0y}t - \frac{1}{2}gt^2$$

Substituting $y = 350$ m for the vertical coordinate of the target, $210\sin(\theta)$ for v_{0x}, $g = 9.81$, and t gives:

$$350 = 210\sin(\theta)\frac{2600}{210\cos(\theta)} - \frac{1}{2}9.81\left(\frac{2600}{210\cos(\theta)}\right)^2$$

or:

$$350 = 2600\tan(\theta) - \frac{1}{2}9.81\left(\frac{2600}{210\cos(\theta)}\right)^2$$

The solution of this equation gives the angle θ at which the projectile has to be fired.

b) A solution of the equation derived in part a obtained by using the `solve` command (in the Command Window) is:

```
>> syms theta
>> Angle = solve('2600*tan(theta) - 0.5*9.81*(2600/(210*cos(theta)))^2 = 350')
Angle =
[ -2.6823398465577220256847788629067]
[ -1.8962381563523770701488298026235]
[  1.2453544972374161683138135806560]
[  .45925280703207121277786452037279]
```

MATLAB displays four solutions. The two positive ones are relevant to the problem.

```
>> Angle1 = Angle(3)*180/pi
```

Converting the solution in the third element of `Angle` from radians to degrees.

```
Angle1 =
224.16380950273491029648644451808/pi
```

MATLAB displays the answer as a symbolic object in terms of π.

```
>> Angle1 = double(Angle1)
Angle1 =
   71.3536
```

Use the `double` command to obtain numerical values for `Angle1`.

>> Angle2 = Angle(4)*180/pi

> Converting the solution in the fourth element of `Angle` from radians to degrees.

Angle2 =
82.665505265772818300015613667102/pi

> MATLAB displays the answer as a symbolic object in terms of π.

>> Angle2 = double(Angle2)

Angle2 =
   26.3132

> Use the `double` command to obtain numerical values for `Angle2`.

c) The solution from part b shows that there are two possible angles and thus two trajectories. In order to make a plot of a trajectory, the x and y coordinates of the projectile are written in terms of t (parametric form):

$$x = v_0 \cos(\theta)t \quad \text{and} \quad y = v_0 \sin(\theta)t - \frac{1}{2}gt^2$$

The domain for t is from $t = 0$ until $t = \dfrac{2600}{210\cos(\theta)}$.

These equations can be used in the `ezplot` command to make the plots as shown in the following program written in a script file.

```
xmax = 2600; v0 = 210; g = 9.81;
theta1 = 1.24535; theta2 = .45925;
t1 = xmax/(v0*cos(theta1));
t2 = xmax/(v0*cos(theta2));
syms t
X1 = v0*cos(theta1)*t;
X2 = v0*cos(theta2)*t;
Y1 = v0*sin(theta1)*t - 0.5*g*t^2;
Y2 = v0*sin(theta2)*t - 0.5*g*t^2;
ezplot(X1,Y1,[0,t1])
hold on
ezplot(X2,Y2,[0,t2])
hold off
```

> Assign the two solutions from part b to `theta1` and `theta2`.

> Plot one trajectory.

> Plot second trajectory.

When this program is executed the following plot is generated in the Figure Window:

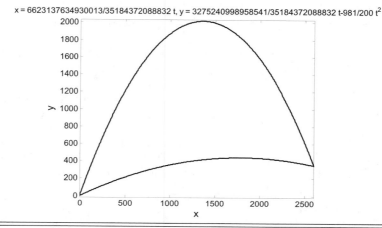

$x = 6623137634930013/35184372088832\ t,\ y = 3275240998958541/35184372088832\ t{-}981/200\ t^2$

Sample Problem 11-3: Bending resistance of a beam

The bending resistance of a rectangular beam of width b and height h is proportional to the beam's moment of inertia I defined by $I = \frac{1}{12}bh^3$. A rectangular beam is cut out of a cylindrical log of radius R. Determine b and h (as a function of R) such that the beam will have maximum I.

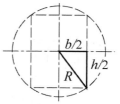

Solution

The problem is solved by following these steps:
a. Write an equation that relates R, h, and b.
b. Derive an expression for I in terms of h.
c. Take the derivative of I with respect to h.
d. Set the derivative equal to zero and solve for h.
e. Determine the corresponding b.

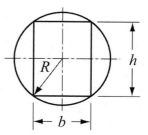

The first step is written by looking at the triangle in the figure. The relationship between R, h, and b is given by the Pythagorean theorem as $\left(\frac{b}{2}\right)^2 + \left(\frac{h}{2}\right)^2 = R^2$.

Solving this equation for b gives $b = \sqrt{4R^2 - h^2}$.
The rest of the steps are done using MATLAB:

```
>> syms b h R
>> b = sqrt(4*R^2 - h^2);          Create a symbolic expression for b.
>> I = b*h^3/12                    Step b: Create a symbolic expression for I.
```

```
I =
1/12*(4*R^2-h^2)^(1/2)*h^3
>> ID = diff(I,h)
ID =
-1/12/(4*R^2-h^2)^(1/2)*h^4+1/4*(4*R^2-h^2)^(1/2)*h^2
>> hs = solve(ID,h)
hs =
[        0]
[        0]
[  3^(1/2)*R]
[ -3^(1/2)*R]
>> bs = subs(b,hs(3))

bs =
(R^2)^(1/2)
>> bss = simple(bs)

bss =
R
```

> MATLAB substitutes *b* in *I*.

> Step *c*: Use the `diff(R)` command to differentiate *I* with respect to *h*.

> The derivative of *I* is displayed.

> Step *d*: Use the `solve` command to solve the equation *ID = 0* for *h*. Assign the answer is to *hs*.

> MATLAB displays four solutions. The positive non-zero solution $\sqrt{3}R$ is relevant to the problem.

> Step *e*: Use the `subs` command to determine *b* by substituting the solution for *h* in the expression for *b*.

> The answer for *b* is displayed. (The answer is *R*, but MATLAB displays $(R^2)^{1/2}$.)

> Use the `simple` command to simplify `bs`.

> The simplified answer for *b* is displayed.

Sample Problem 11-4: Fuel level in a tank

The horizontal cylindrical tank shown is used to store fuel. The tank has a diameter of 6 m and is 8 m long. The amount of fuel in the tank can be estimated by looking at the level of the fuel through a narrow vertical glass window at the front of the tank. A scale that is marked next to the window shows the level of the fuel corresponding to 40, 60, 80, 120, and 160 thousand litters. Determine the vertical position (measured from the ground) of the lines of the scale.

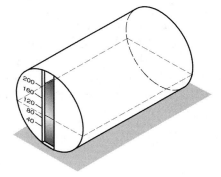

Solution

The relationship between the level of the fuel and its volume can written in the form of a definite integral. Once the integration is carried out, an equation is obtained for the volume in terms of the fuel's height. The height corresponding to a specific volume can then be determined from solving the equation for the height.

The volume of the fuel V can be determined by multiplying the area of the cross section of the fuel A (the shaded area) by the length of the tank L. The cross sectional area can be calculated by integration.

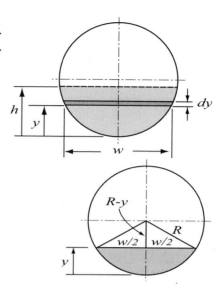

$$V = AL = L \int_0^h w\,dy$$

The width w of the top surface of the fuel can be written as a function of y. From the triangle in the figure on the right, the variables y, w, and R are related by:

$$\left(\frac{w}{2}\right)^2 + (R-y)^2 = R^2$$

Solving this equation for w gives:

$$w = 2\sqrt{R^2 - (R-y)^2}$$

The volume of the fuel at height h can now be calculated by substituting w in the integral in the equation for the volume and carrying out the integration. The result is an equation that gives the volume V as a function of h. The value of h for a given V is obtained by solving the equation for h. In the present problem values of h have to be determined for volumes of 40, 60, 80, 120, and 160 thousand liters. The solution is given in the following MATLAB program (script file):

```
R = 3; L = 8;

syms w y h

w = 2*sqrt(R^2 - (R-y)^2)      Create a symbolic expression for w.

S = L*w              Create the expression that will be integrated.

V = int(S,y,0,h)     Use the int command to integrate S from 0 to h.
                     The result gives V as a function of h.

Vscale = [40:40:200]   Create a vector with the values of V in the scale.

for i = 1:5            Each pass in the loop solves h for one value of V.

   Veq = V-Vscale(i);    Create the equation for h that has to be solved.

   h_ans(i) = solve(Veq);    Use the solve command to solve for h.

end          h_ans is a vector (symbolic with numbers) with the values
             of h that correspond to the values of V in the vector Vscale.

h_scale = double(h_ans)    Use the double command to obtain numerical
                           values for the elements of vector h_ans.
```

When the script file is executed, the outcome from commands that don't have a semicolon at the end are displayed. The display in the Command Window is:

>> w =
2*(6*y-y^2)^(1/2)
— The symbolic expression for w is displayed.

S =
16*(6*y-y^2)^(1/2)
— S is the expression that will be integrated.

V =
8*(6*h-h^2)^(1/2)*h-24*(6*h-h^2)^(1/2)+72*asin(-1+1/3*h)+36*pi
— The result from the integration; V as a function of h.

Vscale =
   40   80   120   160   200
— The values of V in the scale are displayed.

h_scale =
   1.3972   2.3042   3.1439   3.9957   4.9608
— The position of the lines in the scale are displayed.

Units: The unit for length in the solution is meter, which correspond to m³ for the volume (1 m³ = 1,000 L).

Sample Problem 11-5: Amount of medication in the body

The amount M of medication present in the body depends on the rate that the medication is consumed by the body and on the rate that the medication enters the body, where the rate at which the medication is consumed is proportional to the amount present in the body. A differential equation for M is:

$$\frac{dM}{dt} = -kM + p$$

where k is the proportionality constant and p is the rate that the medication is injected into the body.

a)   Determine k if the half-life of the medication is 3 hours.
b)   A patient is admitted to a hospital and the medication is given to him at a rate of 50 mg per hour. (Initially there is no medication in the patient's body.) Derive an expression for M as a function of time.
c)   Plot M as a function of time for the first 24 hours.

Solution

a)   The proportionality constant can be determined from considering the case in which the medication is consumed by the body and no new medication is given. In this case the differential equation is:

$$\frac{dM}{dt} = -kM$$

The equation can be solved with the initial condition $M = M_0$ at $t = 0$:

```
>> syms M M0 k t
>> Mt = dsolve('DM = -k*M','M(0) = M0')
Mt =
M0*exp(-k*t)
```

Use the `dsolve` command to solve $\dfrac{dM}{dt} = -kM$.

The solution gives *M* as a function of time:

$$M(t) = M_0 e^{-kt}$$

Half-life of 3 hours means that at $t = 3$ hours $M(t) = \dfrac{1}{2}M_0$. Substituting this information in the solution gives $0.5 = e^{-3k}$, and the constant *k* is determined from solving this equation:

```
ks = solve('0.5 = exp(-k*3)')
ks =
.23104906018664843647241070715273
```

Use the `solve` command to solve $0.5 = e^{-3k}$.

b) For this part the differential equation for *M* is:

$$\frac{dM}{dt} = -kM + p$$

The constant *k* is known from part *a*, and $p = 50$ mg/h is given. The initial condition is that in the beginning there is no medication in the patient's body, or $M = 0$ at $t = 0$. The solution of this equation with MATLAB is:

```
>> syms p
>> Mtb = dsolve('DM = -k*M + p','M(0)=0')
Mtb =
p/k-p/k*exp(-k*t)
```

Use the `dsolve` command to solve $\dfrac{dM}{dt} = -kM + p$.

c) A plot of `Mtb` as a function of time for $0 \le t \le 24$ can be done by using the `ezplot` command:

```
>> pgiven=50;
>> Mtt = subs(Mtb,{p,k},{pgiven,ks})
Mtt =
216.404-216.404*exp(-.231049*t)
>> ezplot(Mtt,[0,24])
```

Substitute numerical values for *p* and *k*.

In the actual display of the last expression that was generated by MATLAB (`Mtt` = ...) the numbers have many more decimal digits than shown above. The numbers were shortened so that they will fit in the page.

The plot that is generated is:

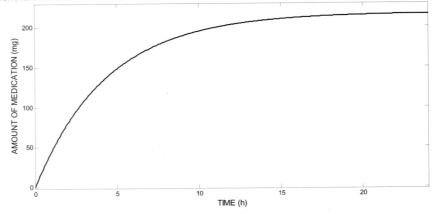

11.10 PROBLEMS

1. Define x as a symbolic variable and create the two symbolic expressions:

$$S_1 = x^3 - 9x^2 + 27x - 27 \text{ and } S_2 = (x+3)^2 - x^2 - 5x - 12$$

Use symbolic operations to determine the simplest form of following expressions:

a) $S_1 \cdot S_2$.

b) $\dfrac{S_1}{S_2}$.

c) $S_1 + S_2$.

d) Use the subs command to evaluate the numerical value of the result from part c for $x = 10$.

2. Define y as a symbolic variable and create the two symbolic expressions:

$$S_1 = (\sqrt{y} + 2)^2 - 2(2\sqrt{y} + 1) \text{ and } S_2 = y^2 - 2y + 4$$

Use symbolic operations to determine the simplest form of following expressions:

a) $S_1 \cdot S_2$.

b) $\dfrac{S_1}{S_2}$.

c) $S_1 + S_2$.

d) Use the subs command to evaluate the numerical value of the result from part c for $x = 5$.

3. Define w as a symbolic variable and create the two symbolic expressions:
$$S_1 = 2w^2 + 6w + 9 \quad \text{and} \quad S_2 = 2w^2 - 6w + 9$$
Use symbolic operations to determine the simplest form of the product $S_1 \cdot S_2$.

4. Define x as a symbolic variable.
   a) Show that the roots of the polynomial:
   $$f(x) = x^5 + 6x^4 - 6x^3 - 64x^2 - 27x + 90$$
   are 1, –2, 3, –3, and –5 by using the `factor` command.
   b) Derive the equation of the polynomial that has the roots: $x = 6$, $x = -4$, $x = -1$, and $x = 2$.

5. Use the commands from Section 11.2 to show that:
   a) $\sin(3x) = 3\sin x - 4\sin^3 x$.
   b) $\sin x \sin y = \dfrac{1}{2}[\cos(x-y) - \cos(x+y)]$
   c) $\begin{aligned} \cos(x+y+z) &= \cos x \cos y \cos z - \sin x \sin y \cos z \\ &\quad - \sin x \cos y \sin z - \cos x \sin y \sin z \end{aligned}$

6. The folium of Descartes is the graph shown in the figure. In parametric form its equation is given by:
   $$x = \frac{3t}{1+t^3} \quad \text{and} \quad y = \frac{3t^2}{1+t^3} \quad \text{for } t \neq -1$$

   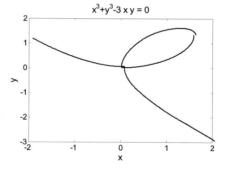

   a) Use MATLAB to show that the equation of the folium of Descartes can also be written as:
   $$x^3 + y^3 = 3xy$$
   b) Make a plot of the folium, for the domain shown in the figure by using the `ezplot` command.

7. A silo in a shape of a cone has a surface area of 280 m², and a height h of 15 m. Determine the radius R of its base. (Write an equation for the surface area in terms of the radius and the height. Solve the equation for the radius, and use the `double` command to obtain a numerical value.)

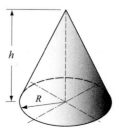

8. A box is manufactured by taking a 20 in. by 30 in. rectangular piece of cardboard, cutting out squares with side x at each corner, and folding up the sides. Determine x such that the volume of the box will be 1,000 in³ (there are two possibilities).

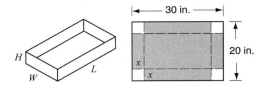

9. A weight W is attached with a ring to two cables that are attached to hinges as shown. The hinge at point A is fixed and the hinge at point B can be translated (without friction) in the horizontal direction. The forces in the cables F_{AC} and F_{BC} depend on the position of hinge B (the distance x) and can be calculated from the equations:

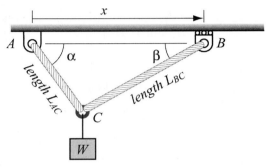

$$F_{AC}\cos\alpha = F_{BC}\cos\beta \quad \text{and} \quad F_{AC}\sin\alpha + F_{BC}\sin\beta = W$$

a) Use MATLAB to derive expressions for the forces F_{AC} and F_{BC} in terms of x, W, and the length of the cables, L_{AC} and L_{BC}.

b) Use the `subs` command to substitute $W = 2000$ N, $L_{AC} = 0.3$ m, and $L_{BC} = 0.5$ m in the expressions that were derived in part a. This will give the forces in the cables as a function of the distance x.

c) Use the `ezplot` command to plot the forces F_{AC} and F_{BC} (both in the same plot) as a function of x, for x starting at 0.4 m and ending close to 0.8 m. What happens when x approaches 0.8 m?

10. An equation of a line in the x-y plane is given by $y = 3x - 2$, and an equation of an ellipse is given by $16x^2 + 32x + 4y^2 - 24y = 52$.

a) Use the `ezplot` command to plot the line and the ellipse in the same plot.

b) Determine the coordinates where the line intersects the ellipse.

11. A tracking radar is locked on an airplane flying at a constant altitude of 5 km, and a constant speed of 540 km/h. The airplane travels along a path that passes exactly above the radar. The radar starts the tracking when the airplane is 100 km away.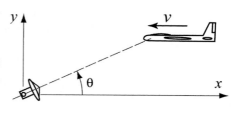

a) Derive an expression for the angle θ of the radar antenna as a function of time.

b) Derive an expression for the angular velocity of the antenna, $\dfrac{d\theta}{dt}$, as a function of time.

c) Make two plots on the same page. One of θ vs. time and the other of $\dfrac{d\theta}{dt}$ vs. time, where the angle is in degrees and the time is in minutes for $0 \le t \le 20$ min.

12. The function $f = 3^x$ has a tangent at point x_P. Derive the equation of the tangent line as a function of x_P. This means the equation of the tangent line will have the form $y = mx + b$, where m and b are functions of x_P. Consider the case where $x_P = 2$. Determine the point at which the tangent intersects the x-axis, and make a plot of f and the tangent line 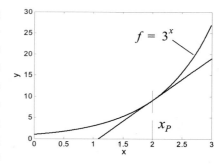 (both in the same figure). Take the domain for the plot of f to be $0 \le x \le 3$. The tangent line should start at the intersection point with the x-axis and end at $x = 3$.

13. Evaluate the indefinite integral $I = \int e^{2x} \sqrt{2 - e^{2x}} \, dx$.

14. The equation of an ellipse is

$$\frac{x^2}{a^2} + \frac{y^2}{b^2} = 1$$

Show that the area A enclosed by the ellipse is given by $A = \pi ab$.

15. A ceramic tile has the design shown in the figure. The shaded area is painted red and the rest of the tile is white. The border line between the red and the white areas follows the equation:

$$y = -kx^2 + 12kx$$

Determine k such that the area of the white and red colors will be the same.

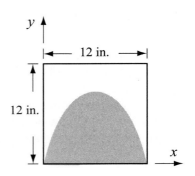

16. Show that the location of the centroid $\bar{x}$ of the half elliptical area that is shown is given by $\bar{x} = \dfrac{4a}{3\pi}$. The coordinate $\bar{x}$ can be calculated from:

$$\bar{x} = \frac{\int x\, dA}{\int dA}$$

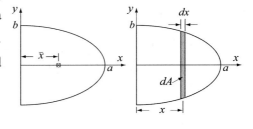

17. The *rms* value of an AC voltage is defined by:

$$v_{rms} = \sqrt{\frac{1}{T}\int_0^T v^2(t')\,dt'}$$

where T is the period of the waveform.

a) A voltage is given by $v(t) = V\cos(\omega t)$. Show that $v_{rms} = \dfrac{V}{\sqrt{2}}$ and is independent of ω. (The relationship between the period T and the radian frequency ω is: $T = \dfrac{2\pi}{\omega}$.)

b) A voltage is given by $v(t) = 2.5\cos(350t) + 3$ V. Determine v_{rms}.

18. Determine the solution of the differential equation:

$$\frac{dy}{dx} = 3y - 1.5yx$$

satisfying the initial condition: $y(0) = 4$. Plot the solution for $0 \le x \le 6$.

19. A resistor R ($R = 0.4\Omega$) and an inductor L ($L = 0.08\,\mathrm{H}$) are connected as shown. Initially, the switch is connected to point A and there is no current in the circuit. At $t = 0$ the switch is moved from A to B, such that the resistor and the inductor are connected to v_S ($v_S = 6\,\mathrm{V}$), and cur-

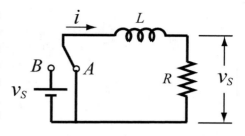

rent starts flowing in the circuit. The switch remains connected to B until the voltage on the resistor reaches 5V. At that time (t_{BA}) the switch is moved back to A.

The current i in the circuit can be calculated from solving the differential equations:

$iR + L\dfrac{di}{dt} = v_S$ During the time from $t = 0$ and until the time when the switch is moved back to A.

$iR + L\dfrac{di}{dt} = 0$ From the time when the switch is moved back to A and on.

The voltage across the resistor v_R at anytime is given by $v_R = iR$.

a) Derive an expression for the current i in terms of R, L, v_S, and t for $0 \le t \le t_{BA}$ by solving the first differential equation.

b) Substitute the values of R, L, and v_S, in the solution of i, and determine the time t_{BA} when the voltage across the resistor reaches 5V.

c) Derive an expression for the current i in terms of R, L, and t, for $t_{BA} \le t$ by solving the second differential equation.

d) Make two plots (on the same page). One for v_R vs. t for $0 \le t \le t_{BA}$, and the other for v_R vs. t for $t_{BA} \le t \le 2t_{BA}$.

20. The velocity of a sky-diver when his parachute is still closed can be modeled by assuming that the air resistance is proportional to the velocity. From Newton's second law of motion the relationship between the mass m of the sky-diver and his velocity v is given by (down is positive):

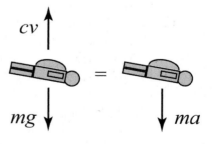

$$mg - cv = m\dfrac{dv}{dt}$$

where c is a drag constant and g is the gravitational constant $g = 9.81\,\mathrm{m/s^2}$.

a) Solve the equation for v in terms of m, g, c, and t, assuming that the initial velocity of the sky-diver is zero.

b) It is observed that 4 s after a 90 kg sky-diver jumps out of an airplane, his velocity is 28 m/s. Determine the constant c.

c) Make a plot of the sky diver velocity as a function of time for $0 \le t \le 30\,\text{s}$.

21. Determine the general solution of the differential equation:

$$\frac{d^2 y}{dx^2} + 3\frac{dy}{dx} - 2y = 0$$

Show that the solution is correct. (Derive the first and second derivatives of the solution, and then substitute back in the equation.)

22. Determine the solution of the following differential equation that satisfies the given initial conditions.

$$\frac{d^2 y}{dx^2} - 4y = 5, \quad y(0) = 0, \quad \left.\frac{dy}{dx}\right|_{x=0} = 1$$

23. Damped free vibrations can be modeled by considering a block of mass m that is attached to a spring and a dashpot as shown. From Newton's second law of motion, the displacement x of the mass as a function of time can be determined by solving the differential equation:

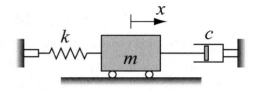

$$m\frac{d^2 x}{dt^2} + c\frac{dx}{dt} + kx = 0$$

where k is the spring constant, and c is the damping coefficient of the dashpot. If the mass is displaced from its equilibrium position and then released, it will start oscillating back and forth. The nature of the oscillations depends on the size of the mass and the values of k and c.

For the system shown in the figure $m = 10\,\text{kg}$, and $k = 28\,\text{N/m}$. At time $t = 0$ the mass is displaced to $x = 0.18\,\text{m}$, and then released from rest. Derive expressions for the displacement x and the velocity v of the mass, as a function of time. Consider the following two cases:

a) $c = 3\,\text{N-s/m}$.

b) $c = 50\,\text{N-s/m}$.

For each case, plot the position x and the velocity v vs. time. (Two plots on one page.) For case *(a)* take $0 \le t \le 20$ s, and for case *(b)* take $0 \le t \le 10$ s.

Appendix:
Summary of Characters,
Commands, and Functions

The following tables list MATLAB's characters, commands, and functions that are covered in the book. The items are grouped by subjects.

Characters and arithmetic operators

Character	Description	Page
+	Addition.	10, 52
−	Subtraction.	10, 52
*	Scalar and array multiplication.	10, 53
.*	Element-by-element multiplication of arrays.	60
/	Right division.	10, 59
\	Left division.	10, 58
./	Element-by-element right division.	60
.\	Element-by-element left division.	60
^	Exponentiation.	10
.^	Element-by-element exponentiation.	60
:	Colon; creates vectors with equally spaced elements, represents range of elements in arrays.	29, 36
=	Assignment operator.	15
( )	Parentheses; sets precedence, encloses input arguments in functions and subscripts of arrays.	10, 34, 35, 144
[ ]	Brackets; forms arrays. encloses output arguments in functions.	28, 29, 31, 144
,	Comma; separates array subscripts and function arguments, separates commands in the same line.	8, 17, 34-37, 144
;	Semicolon; suppresses display, ends row in array.	9, 31
'	Single quote; matrix transpose, creates string.	33, 46-48
...	Ellipsis; continuation of line.	9
%	Percent; denotes a comment, specifies output format.	9

Relational and logical operators

Character	Description	Page
<	Less than.	166
>	Greater than.	166
<=	Less than or equal.	166
>=	Greater than or equal.	166
==	Equal.	166
~=	Not equal.	166
&	Logical AND.	169
|	Logical OR.	169

Relational and logical operators (Continued)

Character	Description	Page
~	Logical NOT.	169

Managing commands

Command	Description	Page
cd	Changes current directory.	79
clc	Clears the Command Window.	10
clear	Removes all variables from the memory.	19
clear x y z	Removes variables x, y, and z from the memory.	19
fclose	Closes a file.	92
fopen	Opens a file.	91
global	Declares global variables.	147
help	Display help for MATLAB functions.	146
lookfor	Search for specified word in all help entries.	145
who	Displays variables currently in the memory.	19, 43
whos	Displays information on variables in the memory.	19, 43

Predefined variables

Variable	Description	Page
ans	Value of last expression.	18
eps	The smallest difference between two numbers.	18
i	$\sqrt{-1}$	18
inf	Infinity.	18
j	Same as i.	18
NaN	Not a number.	18
pi	The number π.	18

Display formats in the Command Window

Command	Description	Page
format bank	Two decimal digits.	12
format compact	Eliminates empty lines.	13
format long	Fixed-point format with 14 decimal digits.	12
format long e	Scientific notation with 15 decimal digits.	12
format long g	Best of 15-digit fixed or floating point.	12
format loose	Adds empty lines.	13
format short	Fixed-point format with 4 decimal digits.	12
format short e	Scientific notation with 4 decimal digits.	12
format short g	Best of 5-digit fixed or floating point.	12

Elementary math functions

Function	Description	Page
abs	Absolute value.	14
exp	Exponential.	14
factorial	The factorial function.	14
log	Natural logarithm.	14
log10	Base 10 logarithm.	14
sqrt	Square root.	14

Trigonometric math functions

Function	Description	Page	Function	Description	Page
acos	Inverse cosine.	14	cos	Cosine.	14
acot	Inverse cotangent.	14	cot	Cotangent.	14
asin	Inverse sine.	14	sin	Sine.	14
atan	Inverse tangent.	14	tan	Tangent.	14

Hyperbolic math functions

Function	Description	Page	Function	Description	Page
cosh	Hyperbolic cosine.	14	sinh	Hyperbolic sine.	14
coth	Hyperbolic cotangent.	14	tanh	Hyperbolic tangent.	14

Rounding

Function	Description	Page
ceil	Round towards infinity.	15
fix	Round towards zero.	15
floor	Round towards minus infinity.	15
rem	Returns the remainder after x is divided by y.	15
round	Round to the nearest integer.	15
sign	Signum function.	15

Creating arrays

Function	Description	Page
diag	Creates a diagonal matrix from a vector. Creates a vector from the diagonal of a matrix.	42
eye	Creates a unit matrix.	32, 56
linspace	Creates equally spaced vector.	30
ones	Creates an array with ones.	32
rand	Creates an array with random numbers.	65, 66
randn	Creates an array with normally distributed numbers.	67
randperm	Creates vector with permutation of integers.	66
zeros	Creates an array with zeros.	32

Handling arrays

Function	Description	Page
length	Number of elements in the vector.	41
reshape	Rearrange a matrix.	42
size	Size of an array.	42

Array functions

Function	Description	Page
cross	Calculates cross product of two vectors.	65
det	Calculates determinant.	58, 65
dot	Calculates scalar product of two vectors.	53, 65
inv	Calculates the inverse of a function.	57, 65
max	Returns maximum value.	64
mean	Calculates mean value.	64
median	Calculates median value.	64
min	Returns minimum value.	64

Array functions (Continued)

Function	Description	Page
sort	Arranges elements in ascending order.	64
std	Calculates standard deviation.	65
sum	Calculates sum of elements.	64

Input and output

Command	Description	Page
disp	Displays output.	84
fprintf	Displays or saves output.	87-94
input	Prompts for user input.	83
uiimport	Starts the Import Wizard	96
xlsread	Imports data from Excel	95
xlswrite	Exports data to Excel	95

Two-dimensional plotting

Command	Description	Page
bar	Creates a vertical bar plot.	125
barh	Creates a horizontal bar plot.	125
fplot	Plots a function.	114
hist	Creates a histogram.	127-129
hold off	Ends hold on.	117
hold on	Keeps current graph open.	117
line	Adds curves to existing plot.	117
loglog	Creates a plot with log scale on both axes.	124
pie	Creates a pie plot.	126
plot	Creates a plot.	108-114
polar	Creates a polar plot.	129
semilogx	Creates a plot with log scale on the x axis.	124
semilogy	Creates a plot with log scale on the y axis.	124
stairs	Creates a stairs plot.	126
stem	Creates a stem plot.	126

Three-dimensional plotting

Command	Description	Page
bar3	Creates a vertical 3-D bar plot.	247
contour	Creates a 2-D contour plot.	246
contour3	Creates a 3-D contour plot.	246
cylinder	Plots a cylinder.	247
mesh	Creates a mesh plot.	243, 244
meshc	Creates a mesh and a contour plot.	245
meshgrid	Creates a grid for a 3-D plot.	241
meshz	Creates a mesh plot with a curtain.	245
pie3	Creates a pie plot.	248
plot3	Creates a plot.	239
scatter3	Creates a scatter plot.	248
sphere	Plots a sphere.	247
stem3	Creates a stem plot	248
surf	Creates a surface plot.	243, 245
surfc	Creates a surface and a contour plot.	245

Three-dimensional plotting (Continued)

Command	Description	Page
surfl	Creates a surface plot with lighting.	246
waterfall	Creates a mesh plot with a waterfall effect.	246

Formatting plots

Command	Description	Page
axis	Sets limits to axes.	122
colormap	Sets color.	243
grid	Adds grid to a plot.	122, 244
gtext	Adds text a plot.	119
legend	Adds legend to a plot.	119
subplot	Creates multiple plots on one page.	130
text	Adds text a plot.	119
title	Adds title to a plot.	119
view	Controls the viewing direction of a 3-D plot.	249
xlabel	Adds label to x axis.	119
ylabel	Adds label to y axis.	119

Math functions (create, evaluate, solve)

Command	Description	Page
feval	Evaluates the value of a math function.	154
fminbnd	Determines the minimum of a function.	264
fzero	Solves an equation with one variable.	262
inline	Creates an inline function.	151

Numerical integration

Function	Description	Page
quad	Integrates a function.	266
quadl	Integrates a function.	267
trapz	Integrates a function.	268

Ordinary differential equation solvers

Command	Description	Page
ode113	Solves a first order ODE.	270
ode15s	Solves a first order ODE.	270
ode23	Solves a first order ODE.	270
ode23s	Solves a first order ODE.	270
ode23t	Solves a first order ODE.	270
ode23tb	Solves a first order ODE.	270
ode45	Solves a first order ODE.	270

Logical Functions

Function	Description	Page
all	Determines if all array elements are nonzero.	172
and	Logical AND.	171
any	Determines if any array elements are nonzero.	172
find	Finds indices of certain elements of a vector.	172
not	Logical NOT.	171

Logical Functions (Continued)

Function	Description	Page
or	Logical OR.	171
xor	Logical exclusive OR.	172

Flow control commands

Command	Description	Page
break	Terminates execution of a loop.	192
continue	Terminates a pass in a loop.	192
else	Conditionally execute commands.	176
elseif	Conditionally execute commands.	178
end	Terminates conditional statements and loops.	174, 180, 183, 187
for	Repeats execution of a group of commands.	183
if	Conditionally execute commands.	174
switch	Switches among several cases based on expression.	180
while	Repeats execution of a group of commands.	187

Polynomial functions

Function	Description	Page
conv	Multiplies polynomials.	211
deconv	Divides polynomials.	211
poly	Determines coefficients of a polynomial.	210
polyder	Determines the derivative of a polynomial.	212
polyval	Calculates the value of a polynomial.	208
roots	Determines the roots of a polynomial.	209

Curve fitting and interpolation

Function	Description	Page
interp1	One-dimensional interpolation.	222
polyfit	Curve fit polynomial to set of points.	215

Symbolic Math

Function	Description	Page
collect	Collects terms in an expression.	290
diff	Differentiates an equation.	300
double	Converts number from symbolic form to numerical form	288
dsolve	Solves an ordinary differential equation.	303
expand	Expands an expression.	291
ezplot	Plots an expression.	306
factor	Factors to product of lower order polynomials.	291
findsym	Displays the symbolic variables in an expression.	289
int	integrates an expression.	301
pretty	Displays expression in math format.	293
simple	Finds a form of an expression with fewest characters.	293
simplify	Simplifies an expression.	292
solve	Solves a single equation, or a system of equations.	294
subs	Substitutes numbers in an expression.	309
sym	Creates symbolic object.	284
syms	Creates symbolic object.	286

Answers to Selected Problems

Chapter 1

2. *a*) 1.7584e+003
   b) 1.0174e+007

4. *a*) 1.4395
   b) 0.2325

6. *a*) 629.1479
   b) 2.0279

8. $r = 4.3718$ in., $A = 240.1759$ in.^{2}

12. *a*) $b = 17.8885$ cm
    b) $\alpha = 31.5881^\circ$

14. $d = 2.5725$

16. *a*) \$1233.82
    b) \$1301.68
    c) \$1302.00

18. 707.95

Chapter 3

2. $y = 1.0e+004$ *
   -0.5954 -0.1000 -0.0116 -0.0008
   -0.0000 0 0.0000 0.0008 0.0116
   0.1000 0.5954 2.7000

4. $z =$ 6.7415 5.4965 5.2579
   5.2421 5.2415

6. $y =$ 2.0000 2.5937 2.7048
   2.7156 2.7169 2.7176 2.7179
   2.7181

8. *a*) $S_n = 0.6883$
   b) $S_n = 0.6926$
   c) $S_n = 0.6931$

12.
5	9957
10	19611
15	28670
20	36857
25	43925
30	49657
35	53881
40	56468
45	57339
50	56468
55	53881
60	49657
65	43925
70	36857
75	28670
80	19611
85	9957

14.
Resistor	Current (A)
R_1	0.6750
R_2	1.1036
R_3	0.8011
R_4	0.3025
R_5	0.9324
R_6	1.2349
R_7	1.9553
R_8	1.3610
R_9	3.3163
R_{10}	2.8877

Chapter 4

2. theta = 37.2750 28.1630
   22.0930 18.0097 15.1373

4.
theta	time (hr)
0	4.0857
10.0000	3.9469
20.0000	3.8407
30.0000	3.7650
40.0000	3.7232
50.0000	3.7276
60.0000	3.8126

333

6.

Time (s)	Distance (m)	Velocity (m/s)
0	0	0
1.0000	0.7750	1.5500
2.0000	3.1000	3.1000
3.0000	6.9750	4.6500
4.0000	12.4000	6.2000
5.0000	19.3750	7.7500
6.0000	27.9000	9.3000
7.0000	37.9750	10.8500
8.0000	49.6000	12.4000
9.0000	62.7750	13.9500
10.0000	77.5000	15.5000

8. $a = 2 \quad b = 3.5$
    $c = 4.2 \quad b = 7$

Chapter 5

2.
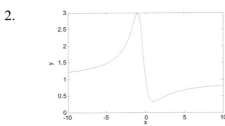

4.
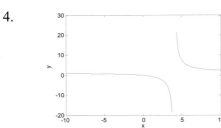

6.
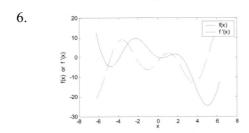

8.
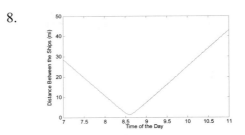

Visibility less than 8 mile from about 8.1 AM until 9 AM.

10.
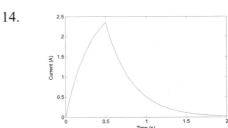

12.
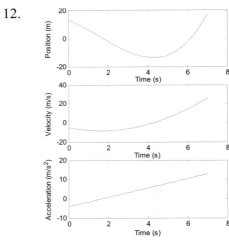

14.
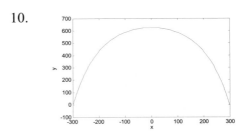

16.

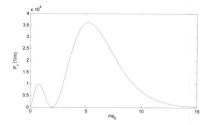

16. *a)*

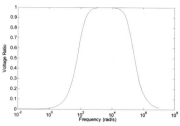

18.

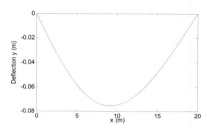

b)

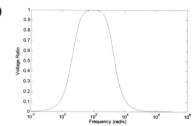

Chapter 6

2. *a)* $y(-3) = -20.1$, $y(5) = 237.5$

b)

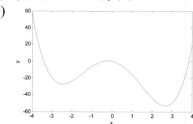

4. *a)* $(3, 21)$
   b) $(1, 2)$

6. 56.5 N-m

8. *a)* -0.7730   0.5797   0.2577
   b) 0.7730   -0.5797   -0.2577
   c) 0.7071   0   -0.7071

10. *a)* 75.6000
    b) 81.6000
        55.8000
        68.8000
        83.4000

14. *a)* 179.1974 MPa -69.1974 MPa
    b) -6.7199 psi -21.2801 psi

Chapter 7

2. *a)* y = 1
   b) y = 0
   c) y = 10

4. y = [2 1 4]

6. *a)* New York 37.6774 °F
      Anchorage 33.1290 °F
   b) New York 17
      Anchorage 13
   c) 11 days, on days: 1   7   9
      14   15   18   19   21   22
      25   26
   d) 1 day, on the 23rd.
   e) 16 days, on days: 7   8   9
      13   14   15   16   17   18
      19   20   23   24   25   26
      27

8. *a)* The equation has two roots
      x1 = 0.345208,
      x2 = -4.345208
   b) The equation has no real roots.
   c) The equation has one root:
      x = -0.333333

18. (11.3099, 15.2971)
    (120.2564, 13.8924)

(207.8973, 19.2354)
(−33.0239, 11.9269)

20.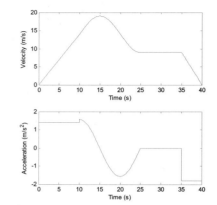

Chapter 8

2. $3x^2 + 7x - 5$

4.   x = 0.0022785 m

8. $a = 7.5$ m, $b = 5$ m

10. *a)*

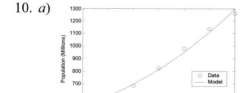

Pop.
1955 = 644.7597 millions

b)

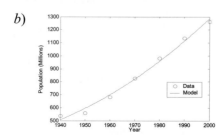

Pop. 1955 = 645.1964 millions

c)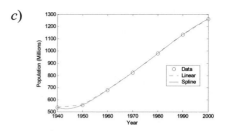

Pop. 1955L = 619.5 millions
Pop. 1955S = 613.058 millions

12.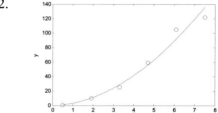

14. *a)* $k = 12.2603$ MPa $\sqrt{\text{mm}}$
$\sigma_0 = 28.2938$ MPa
$\sigma_y = 83.1237$ MPa for $d = 0.05$
mm

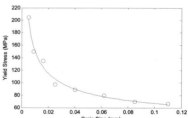

b) $\sigma_y = 84.9091$ MPa for $d = 0.05$
mm

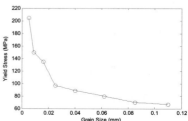

c) $\sigma_y = 88.5457$ MPa for $d = 0.05$
mm

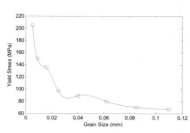

16 $C = 1.5682\text{e-}6 \; \dfrac{\text{Kg}}{\text{m} \cdot \text{s} \cdot \text{K}^{1/2}}$

$S = 148.1622 \text{ K}$

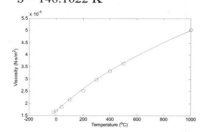

Chapter 9

2.

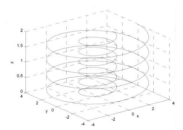

4.

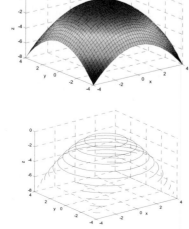

6.

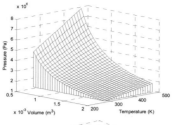

8   *a)*

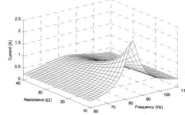

b)

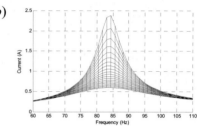

$1/(2\pi\sqrt{LC}) = 83.882 \text{ Hz}$

10.

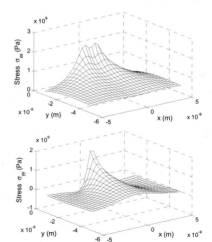

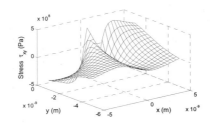

Chapter 10

2.   3.4664, 6.2869, 9.1585

4.   53.1653^o

6.   24.2280^o , 80.5135 N

8. r = 12.2474 cm, h = 17.3205 cm

10. 0.8774

12. E = 6.0986e+006 N/C

14. 264000 m^2

16.

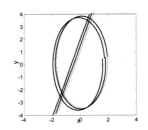

18.

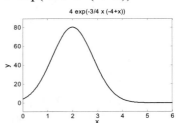

Chapter 11

2.   *a*)   y^3+8
     b)   (y+2)/(y^2-2*y+4)
     c)   -y+6+y^2
     d)   26

4.   *a*)   (x-1)*(x-3)*(x+5)*(x+3)*(x+2)
     b)   x^4-3*x^3-24*x^2+28*x+48

8.   5 in.,   2.9289 in.

10. *a*)

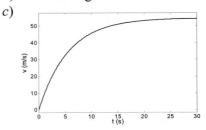

     b) $x = (5/13)\pm(1/13)\sqrt{142}$
           $y = (-11/13)\pm(3/13)\sqrt{142}$

12. (-1+2*log(3))/log(3)

18. y = 4*exp(-3/4*x*(-4+x))

20. *a*)   g/c*m-exp(-c/m*t)*g/c*m
     b)   16.1489 kg/s
     c)

Index